MATHEMATICS FOR THE FUNDAMENTALS OF ENGINEERING (EIT) EXAMINATION

Other McGraw-Hill Books of Interest

Chopey • HANDBOOK OF CHEMICAL ENGINEERING CALCULATIONS
Gieck & Gieck • ENGINEERING FORMULAS
Hicks • STANDARD HANDBOOK OF ENGINEERING CALCULATIONS
Kurtz • CALCULATIONS FOR ENGINEERING ECONOMIC ANALYSIS
Kurtz • HANDBOOK OF APPLIED MATHEMATICS FOR ENGINEERS AND SCIENTISTS
Seidman, et al • HANDBOOK OF ELECTRIC POWER CALCULATIONS
Tuma • ENGINEERING MATHEMATICS HANDBOOK
Wadsworth • HANDBOOK OF STATISTICAL METHODS FOR ENGINEERS AND SCIENTISTS

MATHEMATICS FOR THE FUNDAMENTALS OF ENGINEERING (EIT) EXAMINATION

Max Kurtz, P.E.

Consulting Engineer and Educator; Author,
Handbook of Applied Mathematics for Engineers
and Scientists, Handbook of Engineering Economics,
Calculations for Engineering Economic Analysis,
Structural Engineering for Professional Engineers'
Examinations, Engineering Economics for
Professional Engineers' Examinations,
Comprehensive Structural Design Guide, Steel
Framing of Hip and Valley Rafters;
Project Editor, Civil Engineering Reference Guide;
Contributing Author, Standard Handbook of
Engineering Calculations

McGRAW-HILL

New York San Francisco Washington, D.C. Auckland Bogotá
Caracas Lisbon London Madrid Mexico City Milan
Montreal New Delhi San Juan Singapore
Sydney Tokyo Toronto

Library of Congress Cataloging-in-Publication Data

Kurtz, Max, (date)
Mathematics for the fundamentals of engineering (EIT) examination / Max Kurtz.

p. cm.
Includes index.
ISBN 0-07-036022-7 (alk. paper)
1. Engineering—United States—Examinations—Study guides. 2. Engineering mathematics—Study guides. 3. Engineering mathematics—Problems, exercises, etc. 4. Engineers—Certification—United States. I. Title.
TA159.K78 1997
620'.0076—DC20 96-41047
CIP

McGraw-Hill

1 2 3 4 5 6 7 8 9 0 DOC/DOC 9 0 1 0 9 8 7 6

ISBN 0-07-036022-7

The sponsoring editor for this book was Harold B. Crawford, the editing supervisor was Peggy Lamb, and the production supervisor was Suzanne W. B. Rapcavage. It was set in Times Roman by Santype Ltd.

Printed and bound by R. R. Donnelley & Sons Company.

This book is printed on recycled, acid-free paper containing a minimum of 50% recycled de-inked fiber.

CONTENTS

Chapter 11. Statistics and Probability of Continuous Variables 211

Chapter 12. Vector Analysis 223

Appendix. Derivatives, Integrals, and Laplace Transforms 233

Index 239

PREFACE

This book is designed for use primarily by the candidate for the professional engineer's (PE) license. Its main objectives are the following:

1. To enable the reader to solve the specific problems in mathematics that are presented in the Fundamentals of Engineering (FE) examination, in both the morning and afternoon sessions. Mathematics is a substantial part of the FE examination, and the examinee who solves the mathematics problems correctly has a solid base on which to build a passing grade.
2. To give the reader the understanding of mathematics and the proficiency in applying it that are essential for mastering each specific area of engineering, such as dynamics, mechanics of materials, fluid mechanics, and electricity. Since all engineering is essentially an application of mathematics, it is mandatory that the candidate for licensure have a firm grasp of mathematics in order to solve problems in engineering.

In addition to the PE candidate, this book will be useful to anyone seeking a general review of mathematics.

The FE examination, which was formerly known as the Engineer-in-Training examination, is a multiple-choice examination of the closed-book type, and the only reference material that the examinee has available is the REFERENCE HANDBOOK issued by the National Council of Examiners for Engineering and Surveying (NCEES). Eventually, the examination will be given exclusively in the SI system of units.

To satisfy the needs of the PE candidate fully, this book has been written for a closed-book examination, and its terminology and notational system coincide fully with those in the NCEES Handbook. Although the system of units in which numerical quantities are expressed is not crucial in mathematics, this book nevertheless applies solely the SI system to assist the reader in becoming fully conversant with that system.

This book explains the definitions, laws, principles, and techniques of mathematics in a simple and readily understandable manner, and it then offers the reader intensive drill in problem solving by presenting the solution to 271 problems that are typical of the FE examination. Each solution is explained in complete detail, every step of the way, thereby enabling the reader to understand the solution fully and easily.

To prepare the examinee thoroughly for the FE examination, this book runs the gamut of material from the basic definitions and laws of algebra to advanced calculus and vector analysis. Its subject matter mirrors the material in mathematics that is recorded in the NCEES Handbook. However, since the examination may include material that goes beyond the Handbook and that an engineer can reasonably be expected to know, this book also includes such subjects as set theory, Markov chains, and the negative-exponential probability distribution.

This book embodies the experience I have acquired in teaching PE review courses continually since 1961 (and other engineering courses since 1959). The PE courses have encompassed mathematics, all the basic engineering sciences, engineering economics, civil engineering, and mechanical engineering. During this period, I have learned how engineers think, visualize, and conceptualize, and I have learned which pedagogical methods are most effective in presenting courses to engineers. Everything I have learned along these lines was woven into the writing of this book.

I am highly grateful to Peggy Lamb of the McGraw-Hill staff for her invaluable services on this project. I am also grateful to my wife, Ruth Ingraham Kurtz, B.E. in E.E., formerly project manager in computer systems development, for her constant assistance in the preparation of the manuscript.

Max Kurtz

INTRODUCTION

This review book for the FE examination has a vast number of unique features that set it apart from all other review books. We wish to highlight several of these features.

Recognizing that the examinee must work rapidly to complete the FE examination on time, this book stresses use of the calculator to perform such operations as solving a system of two or three simultaneous equations, solving a quadratic equation where the roots are real numbers, and transforming the rectangular coordinates of a point to the corresponding polar coordinates, and vice versa. These advantages of the calculator are of particular importance in solving problems in electricity. For example, the analysis of a dc circuit often gives rise to a system of three simultaneous equations in which the unknown quantities are the mesh currents. The older method of solving the system of equations by use of determinants is far too cumbersome to be feasible in the FE examination. By contrast, the calculator yields the solution within seconds. Again recognizing the time constraint under which the FE examinee must operate, this book places emphasis on solving a problem by the simplest, least time-consuming method that can be devised.

This book uses an intensely *practical* approach to mathematics in order to make the material meaningful, interesting, and readily understandable. It is written in a down-to-earth rather than an abstruse manner, and it applies simple logic to the fullest extent possible. This practical approach is illustrated by the solution to Example 5.25, which involves probability. In the conventional treatment, a problem of this type is solved by applying Bayes' theorem. This book, by contrast, solves Example 5.25 very simply by equating the probability of an event to its relative frequency in the long run. This practical approach to probability enables the reader to solve problems of this type speedily and without the need to memorize an abstract law or to fathom its meaning.

Definitions constitute the foundation of every subject, and this book accordingly places great emphasis on defining every term in mathematics in the simplest, clearest, and most meaningful manner possible. The definition of a random variable and of probability in Art. 5.1, of a probability distribution in Art. 5.6, and of a frequency curve in Art. 11.1.1 serve as illustrations. Other review books "define" the various types of mean (arithmetic, geometric, harmonic, and root mean square) merely by presenting a formula for calculating the mean. This book, by contrast, presents a very simple definition of a mean *in general*, and it then proceeds to show what each specific type of mean *signifies*. Thus, the reader learns, not simply how to calculate a particular mean, but why and when that mean is applied. Moreover, after presenting a definition, this book often follows it immediately with a numerical example that illustrates the definition and implants it firmly in the reader's mind.

Because they lend themselves to the multiple-choice type of examination, problems pertaining to permutations and combinations and to probability apppear frequently in the FE examihnation. This book develops these subjects by applying simple logic, and it demonstrates how to solve problems of a nonstandard type. For example, Art. 1.25 presents the solution to problems in which it is necessary to form a permutation or combination when certain restrictions apply, and Examples 5.3 and 5.4 demonstrate how to calculate the probability of an event that is extraordinary in some respect. The FE examination contains nonstandard problems of this type, and they cannot be solved merely by applying some set equation. They can be solved very simply by forming a satisfactory permutation or combination. The principle that underlies the study of permutations is the *law of multiplication*, and this law does not even appear in other review books.

This book uses diagrams extensively to stimulate the reader's interest and to make the material more meaningful and more easily remembered. For example, it applies Venn diagrams in the study of probability, it presents Fig. 3.1 to establish effectively the meaning of the minor and cofactor of a matrix element, and it presents Fig. 8.17 to demonstrate the meaning of partial derivatives of both the first and second order. This book contains precisely 100 illustrations.

From the viewpoint of the engineer, one of the most important subjects in mathematics is that pertaining to the properties of areas, volumes, and masses, for this subject is basic to the study of mechanics of materials, fluid mechanics, and dynamics. This subject tends to be treated superficially in other review books, but it is explored thoroughly and intensively in this book.

Other subjects in mathematics that are crucial to engineering also receive thorough coverage in this book. Because the analysis of ac circuits is based on the graphical representation of complex numbers, this book places strong emphasis on that subject. Similarly, because certain basic principles of plane geometry are applied recurrently in engineering, this book presents an exposition of those principles in Chap. 6. Vector analysis, which is applied extensively in dynamics, receives exhaustive treatment in Chap. 12. Fourier series, which are applied in electricity to obtain nonsinusoidal voltage waveforms, are studied in Chap. 9. Matrix algebra, which has very numerous applications in engineering, is explored in depth in Chap. 3.

This book freely applies engineering principles where they offer the simplest possible method of solution to problems in mathematics. Example 8.25 serves as an illustration. In this example, a ladder rests against the wall of a building and is slipping outward at a specified rate. It is necessary to compute the rate at which the upper end of the ladder is descending. This example is first solved by the conventional method of calculus, but it is then solved by a far simpler method: locating the instantaneous center of rotation of the ladder. In this situation, a concept that is applied in dynamics allows us to circumvent mathematical complexity.

In many instances, the solution to an example in this book is subjected to verification, either precise or approximate. To be sure, the FE examination cannot require the examinee to verify a result. However, these verifications serve an important function: to give the reader a deeper understanding of a calculated result and to stimulate the reader's interest. Thus, the solutions to Examples 8.28, 8.29, and 8.30, which are obtained by applying l'Hôpital's rule, are verified approximately but with a very high degree of precision by use of the calculator. These verifications make the solutions to these examples far more meaningful to the reader, for they transform a seemingly abstract principle to a very practical reality. Similarly, the solution to Example 8.25, which involves related rates of change, is verified in an approximate but highly precise manner. A verification of this type gives engineers confidence that their calculated results are correct.

This book is self-contained, and a system of cross references ties all related material together. As a result, it is possible for the reader to review a particular subject if necessary when that subject is applied at a later point in the book.

RECOMMENDATIONS TO THE READER

The recommendations that we shall offer fall into three categories: those that pertain to the study of mathematics in general, those that pertain to the study of this book for the specific purpose of preparing for and passing the FE examination, and those that pertain to the process of taking the examination.

Studying Mathematics in General

The study of mathematics can be made far more effective and even fascinating if we develop certain habits of thought and follow certain procedures.

One such habit is to rely more on visualization than on rigid rules that must be memorized or found by referring to some book. As an illustration, assume that we have located a turning point of a curve. By what criterion do we determine whether it is a local maximum or minimum point? The NCEES Handbook states that it is a maximum point if the second derivative is negative at that point and it is a minimum point if the second derivative is positive. However, rather than refer to the Handbook for this information, we ourselves can deduce it instantly. We simply visualize a curve with a local maximum point. As we pass this point in moving from left to right, the slope of the tangent changes from positive to negative; therefore, the second derivative is negative.

In trigonometry, our study of the solution of a triangle becomes both intriguing and illuminating if we actually *construct* the specified triangle with ruler, protractor, and compasses. Thus, by constructing Fig. 6.12, we discover that the solution to Example 6.8 is ambiguous because the construction yields two triangles rather than one. This construction also reveals the conditions under which an example of this type has a unique solution or no solution whatever.

Our study of mathematics is enriched if we view every mathematical principle and relationship in terms of its potential practical applications. The trammel method of constructing an ellipse of specified size, which is presented in Example 7.15, is illustrative in this respect. The ellipsograph, which is a device that generates an ellipse by the continuous motion of a point, is based on this method. Thus, a very simple principle of trigonometry proved to be of vast practical importance. You will also find it enjoyable and instructive to construct an ellipse by the trammel method. Simply take a straightedge, mark three points on it, move the straightedge in the prescribed manner, and mark successive positions of the generating point. You will find that the curve thus formed is indeed an ellipse, and you will see how the shape of the ellipse varies as you change the relative positions of the points. You yourself can apply the trammel method to shape an elliptical garden or any other elliptical object.

The study of mathematics also becomes far more meaningful and productive if we seek to understand the *significance* of each definition and expression. When we are given a set of values of a variable, what do the geometric mean and root mean square of those values tell us?

Similarly, it is very helpful if we understand our *motive* in defining certain terms. Thus, we define the moment of inertia of a body because this physical property governs the behavior of the body when it is subjected to a moment. Our study of complex numbers and their graphical representation becomes more stimulating when we realize that complex numbers are applied in electricity in the analysis of ac circuits for a very simple reason: The use of complex numbers facilitates our calculations.

Above all, it is imperative that we approach each subject in mathematics with the realization that it is understandable and of practical significance, not mysterious and unfathomable. The subject of probability illustrates this point. Many books present this subject in a highly esoteric manner, thereby confusing and intimidating the reader and effectively concealing the fact that the subject is really simple and logical. In this book, the subject of probability is developed in a manner that reveals its innate simplicity.

Studying This Book as a Means of Preparing for the FE Examination

The first recommendation we can offer is that you study this book thoroughly and intensively and that you perform each calculation that is part of the solution of an example. Try to solve each example yourself before proceeding to the solution presented in this book. As you solve examples that typify the FE examination problems, you will acquire the proficiency and speed that are required for passing the examination.

Since the only reference book you will have available while taking the examination is the NCEES Handbook, it is imperative that you become thoroughly familiar with both the contents of the Handbook and its arrangement of material. For example, if you must apply the equation for the curvature of a curve in solving an examination problem, you should be able to locate this equation in the Handbook within seconds. The Handbook is issued by the state licensing boards. If you do not possess one at present, obtain a copy. Familiarity with the NCEES Handbook can be acquired most effectively in this manner: As you study a particular subject in the present book, refer to the corresponding section in the NCEES Handbook.

The terminology and notational system in the present book conform with those in the Handbook, but they may differ in some instances from the terminology and notational system that was used in your engineering courses at school. To illustrate this point, we shall cite a topic in dynamics. In many engineering books, the turning effect of a force on a body is called a *torque*, and it is denoted by T. In the NCEES Handbook, the turning effect is called a *moment*, and it is denoted by M. Therefore, you must become fully familiar with the terminology and notational system of the Handbook.

If you are not accustomed to working with the SI system of units, it is essential that you overcome this handicap. Books that explain SI are available in every engineering library. The NCEES Handbook presents a table of the prefixes that are used in SI to denote multiples and submultiples of the base units; it appears on page 1. In performing calculations in SI, it is advisable to apply the base units (meter, kilogram, second) and then adjust the calculated result by affixing the appropriate prefix.

The scientific calculator has replaced the slide rule as the basic working tool of the engineer. Become thoroughly familiar with the special functions of the calculator that are of particular importance in the FE examination, such as those applied in solving a system of three simultaneous equations, converting the rectangular coordinates of a point to the corresponding polar coordinates, and solving a quadratic equation with real roots. These functions are fully utilized in solving numerous examples in this book; thus, following these solutions induces familiarity with these functions.

A word of caution concerning use of the calculator is in order, for a calculator does have certain limitations. This is true, for example, with reference to inverse trigonometric functions. The calculator offers only one solution when in reality two solutions exist. Assume as an illustration that our calculations in solving a problem lead to the intermediate result $\sin \theta = -0.40$ and that we must find θ. The calculator presents θ as an angle that lies in the fourth quadrant. However, there is also an angle in the third quadrant that satisfies this equation, and the latter angle may be the one that is valid in a given situation. Manifestly, a result obtained by calculator cannot be accepted blindly.

Taking the Examination

It is advisable to visit the examination site beforehand to estimate the travel time and to become familiar with the parking facilities. The following materials must be brought to the examination:

1. The admission card, with a photograph of yourself.
2. A calculator of the silent type. You should learn whether the board of examiners in your state allows use of a programmable calculator. Since a calculator may stop functioning during the examination, bring two calculators if possible, or bring additional batteries.
3. A watch. As you proceed through the examination, it is imperative that you remain time-conscious to ensure that you complete the examination within the allotted time.
4. An adequate supply of pencils (of the type specified by the examiners) and of erasers. You may have occasion to change an answer.
5. Lunch and bottled water. If you find that chewing gum tends to relax you, bring gum. Smoking is not permitted in the examination room.

Engineers generally find that the most efficient procedure in taking the examination consists of devoting their efforts first to the subjects with which they are most familiar and then turning to the remaining subjects. Again, the FE examination is multiple-choice, and no penalty is incurred if an answer is incorrect. However, even if you guess an answer, do not guess blindly; try to determine which of the possible answers seems most logical.

In selecting a multiple-choice answer, it is essential to avoid falling into a trap occasionally devised by the examiners. For example, assume that we are required to find both the magnitude and direction of a force. Two of the possible answers may have the correct magnitude and inclination of the force, but one has the incorrect *sense*. In the haste of the examination, there is a danger of inadvertently selecting the incorrect answer because its appearance is so similar to the correct one.

In a multiple-choice examination, the correct answer to a mathematics problem can sometimes be identified most quickly by a trial-and-error method in which we *test* each possible answer. As an illustration, assume that we are required to solve a second-order differential equation. We can take the first expression that is offered, write the first and second derivatives, and then substitute in the given equation to determine whether the equation is satisfied. If it is not, we proceed to the second expression and repeat the test, continuing in this manner until the correct expression is found.

It occasionally happens that the statement of an examination problem contains a superfluous item of information. For example, the mass of a body may be given even though it is not relevant. In solving a problem, do not be unduly concerned if you find that you have not applied some given value; it may be superfluous.

Problem statements in engineering economics are sometimes quite lengthy, and the information may not be presented in the most logical sequence. It is advisable to read the problem once without paying close attention to the numerical values; our objective in the first reading is simply to establish the *nature* of the problem. We can then read the problem a second or even a third time to assimilate fully the numerical values.

I wish you success!

MATHEMATICS FOR THE FUNDAMENTALS OF ENGINEERING (EIT) EXAMINATION

CHAPTER 1
ALGEBRA

1.1 CLASSIFICATION OF NUMBERS

An *integer* is a whole number, and it can be positive, negative, or 0. A number that can be expressed in the form p/q, where p and q are integers and $q \neq 0$, is said to be *rational*; a number that cannot be expressed in this form is called *irrational*. For example, it can be demonstrated that the square root of any positive integer that is not a perfect square, such as 8, is irrational. Since we may set $q = 1$, all integers are rational.

The square root of a negative number is described as *imaginary*; for contradistinction, all other numbers are described as *real*. A number that is formed by combining a real and an imaginary number with a plus or minus sign is called a *complex number*. Thus, $-9 + \sqrt{-2}$ and $3 - \sqrt{-7}$ are complex numbers.

Let x denote a real number. The *absolute* (or *numerical*) *value* of x is denoted by $|x|$. The absolute value of a nonnegative number is the number itself; the absolute value of a negative number is the corresponding positive number. Thus,

$$|9| = 9 \qquad |0| = 0 \qquad |-3| = 3$$

Let a, b, and c denote integers such that $a = bc$. The numbers b and c are called *factors* or *divisors* of a, and a is said to be a *multiple* of b and of c. For example, 8 is a factor of 32, and 32 is a multiple of 8. A positive integer that has factors in addition to 1 and itself is described as a *composite number*, and one that lacks such factors is called a *prime number*. By convention, 1 is not considered to be a prime number, and the first six prime numbers are 2, 3, 5, 7, 11, and 13.

Consider an equation of the form

$$a_0 x^n + a_1 x^{n-1} + a_2 x^{n-2} + \cdots + a_n = 0$$

where the a's are all integers and n is a positive integer. A number that satisfies such an equation is called an *algebraic number*, and one that fails to do so is called a *transcendental number*. An algebraic number can be real or complex. For example, $3 + \sqrt{-7}$ is an algebraic number because it satisfies the equation $x^2 - 6x + 16 = 0$. Similarly, $\sqrt{5}$ is an algebraic number because it satisfies the equation $x^2 - 5 = 0$. On the other hand, it is known that π is a transcendental number. It follows at once that every rational number is an algebraic number. For example, 9/17 is an algebraic number because it satisfies the equation $17x - 9 = 0$.

1.2 TYPES OF VARIABLES

Let x denote a variable, x_i denote any value that x can assume, and x_j denote the next higher value of x. Now set $d = x_j - x_i$. If d has a specific value, x is a *discrete* or *step* variable. On the other hand, if d can be less than any number we may possibly specify, x is a *continuous* variable. For example, the number of neutrons in a potassium atom is a discrete variable (20, 21, or 22), and the distance between a moving point and a fixed reference point is a continuous variable.

1.3 SIGNIFICANT FIGURES

When a number is expressed in truncated form, it is said to be *rounded.* In rounding a number, we must determine in advance how many significant figures are to be used. The set of *significant figures* in a number extends from the first nonzero digit at the left to the last recorded digit at the right. For example, the number 81.03 has four significant figures, the number 0.00378 has three significant figures, and the number 823.70 has five significant figures.

1.4 DEFINITIONS PERTAINING TO SUBTRACTION AND DIVISION

Let a and b denote real numbers. In the operation $a - b$, the quantities a and b are called the *minuend* and *subtrahend*, respectively.

Now let c and d denote two positive integers such that $c > d$ and c is not a multiple of d. Let qd denote the largest integer that is a multiple of d and less than c, and let $r = c - qd$. When we divide c by d, q is the *quotient* and r is the *remainder.* For example, when we divide 37 by 8, $qd = 4 \times 8 = 32$, and $r = 37 - 32 = 5$. Thus, the quotient is 4 and the remainder is 5.

1.5 DEFINITIONS PERTAINING TO EXPRESSIONS

When numbers (or letters that represent numbers) are connected by arithmetic operations, they constitute an *expression.* If an expression contains solely multiplication, division, and exponentiation, either singly or in combination, it is called a *term.* Thus, $8x^3/y^2$ is a term. When terms are combined by addition or subtraction (or both), they constitute a *multinomial.* Thus, $2xy^4 + 7x^{1.8} - 6y^2 + 5z$ is a multinomial. A *binomial* and *trinomial* are multinomials consisting of two and three terms, respectively.

If all exponents in a multinomial are positive integers, the expression is a *polynomial.* For example, $7xy^5 + 9x^3 - 3y^2$ is a polynomial. Assume that a polynomial contains the term $ax^p y^q z^r$, where x, y, and z are variables and a is a constant. The *degree* of this term is the sum of the exponents of the variables, or $p + q + r$. The degree of a polynomial equals the highest degree of its terms.

EXAMPLE 1.1 Determine the degree of the polynomial $8x^3yz - 13xy^5 - 2y^2z^6$.

SOLUTION The first term is of degree 5, the second term is of degree 6, and the third term is of degree 8. Therefore, the polynomial is of degree 8.

1.6 LAWS OF ALGEBRAIC OPERATIONS

We now present the basic laws of algebra.

Addition and Multiplication. The laws pertaining to addition and multiplication are as follows:

1. Addition and multiplication are *commutative*. Thus,

$$a + b = b + a \qquad \text{and} \qquad ab = ba$$

2. Addition and multiplication are *associative*. Thus,

$$(a + b) + c = a + (b + c) \qquad \text{and} \qquad (ab)c = a(bc)$$

3. Multiplication is *distributive* with respect to addition and subtraction. Thus,

$$a(b + c) = ab + ac \qquad \text{and} \qquad a(b - c) = ab - ac$$

By an extension of the distributive law, we have

$$(a + b)(c + d) = ac + ad + bc + bd$$

Factoring. When a given polynomial is transformed to an expression that is the product of several quantities, the polynomial is said to be *factored*. We shall consider a particular type of factoring. We have the following:

$$(x + a)(x + b) = x^2 + (a + b)x + ab$$

Therefore, to factor a trinomial of the form $x^2 + cx + d$, we seek two numbers that have a sum of c and a product of d.

EXAMPLE 1.2 Factor each of the following trinomials:

$$x^2 - 7x + 12 \qquad x^2 + 8x - 20$$

SOLUTION With reference to the first trinomial, the numbers -4 and -3 have a sum of -7 and a product of 12. With reference to the second trinomial, the numbers 10 and -2 have a sum of 8 and a product of -20. Therefore,

$$x^2 - 7x + 12 = (x - 4)(x - 3)$$

$$x^2 + 8x - 20 = (x + 10)(x - 2)$$

The following polynomials arise frequently:

$$x^2 + 2xy + y^2 = (x + y)^2$$

$$x^2 - y^2 = (x + y)(x - y)$$

Negative Numbers. Let a and b denote positive numbers; then $-a$ and $-b$ are negative numbers. The laws governing operations with negative numbers are as follows:

$$a + (-b) = a - b \qquad a - (-b) = a + b$$

$$a(-b) = -ab \qquad (-a)(-b) = ab$$

$$\frac{a}{-b} = \frac{-a}{b} = -\frac{a}{b} \qquad \frac{-a}{-b} = \frac{a}{b}$$

Now let c and d denote any real numbers. The number c/d is positive if c and d are both positive or both negative; otherwise, c/d is negative.

Fractions. The laws governing the addition, multiplication, and divisions of fractions are as follows:

$$\frac{a}{b} + \frac{c}{b} = \frac{a+c}{b}$$

$$\frac{a}{b}\frac{c}{d} = \frac{ac}{bd} \qquad \frac{a/b}{c/d} = \frac{a}{b}\frac{d}{c} = \frac{ad}{bc}$$

The numerator and denominator of a fraction can be multiplied or divided by the same number without changing the value of the fraction. Expressed symbolically,

$$\frac{a}{b} = \frac{ac}{bc} = \frac{a/c}{b/c}$$

If the denominator of a fraction is an irrational number, it is often possible to transform the fraction to one in which the denominator is rational. The process is called *rationalizing the denominator*. We shall illustrate the procedure.

EXAMPLE 1.3 Rationalize the denominator in $a/(b + \sqrt{c})$.

SOLUTION

$$\frac{a}{b + \sqrt{c}} = \frac{a}{b + \sqrt{c}}\,\frac{b - \sqrt{c}}{b - \sqrt{c}} = \frac{a(b - \sqrt{c})}{b^2 - c}$$

A fraction is said to be in its *lowest terms* if the numerator and denominator contain no common factor except 1. Therefore, a given fraction can be reduced to its lowest terms by dividing its numerator and denominator by their common factors.

EXAMPLE 1.4 Reduce each of the following fractions to its lowest terms:

$$\frac{9a - 12b}{6c} \qquad \frac{8x + 8y}{4x^2 - 4y^2} \qquad \frac{x^4 + 8x^3 + 15x^2}{x + 3}$$

SOLUTION

$$\frac{9a - 12b}{6c} = \frac{3(3a - 4b)}{3(2c)} = \frac{3a - 4b}{2c}$$

$$\frac{8x + 8y}{4x^2 - 4y^2} = \frac{8(x + y)}{4(x^2 - y^2)} = \frac{8(x + y)}{4(x + y)(x - y)} = \frac{2}{x - y}$$

$$\frac{x^4 + 8x^3 + 15x^2}{x + 3} = \frac{x^2(x^2 + 8x + 15)}{x + 3} = \frac{x^2(x + 5)(x + 3)}{x + 3} = x^3 + 5x^2$$

Consider the equation

$$\frac{a}{b} = \frac{c}{d} \tag{a}$$

If we multiply both sides of the equation by bd, we obtain

$$ad = bc \tag{b}$$

Equation (b) is said to be obtained from Eq. (a) by *cross-multiplying* the fractions in Eq. (a).

Parenthetical Expressions. The plus and minus signs are referred to collectively as *algebraic signs.* When an algebraic sign is omitted, the plus sign is understood. If an expression in parentheses is to be added to or subtracted from some quantity, the parentheses may be removed in accordance with this rule: Retain the algebraic signs within the expression if the expression is to be added; change the algebraic signs within the expression if the expression is to be subtracted. Thus,

$$a + (b - c - d + e - f) = a + b - c - d + e - f$$

$$a - (b - c - d + e - f) = a - b + c + d - e + f$$

1.7 EXPONENTS

Let m and n denote positive integers. The notation a^n denotes a number having a as a factor n times. Thus,

$$3^5 = 3 \cdot 3 \cdot 3 \cdot 3 \cdot 3 = 243$$

In the expression a^n, the number n is called the *exponent* or *power* of a.

From the definition of an exponent, it follows that

$$a^m a^n = a^{m+n} \tag{1.1}$$

$$(a^m)^n = (a^n)^m = a^{mn} \tag{1.2}$$

$$(ab)^n = a^n b^n \tag{1.3}$$

$$\left(\frac{a}{b}\right)^n = \frac{a^n}{b^n} \tag{1.4}$$

Assume that $m > n$. Then

$$\frac{a^m}{a^n} = a^{m-n} \tag{1.5}$$

By extension, the foregoing laws are assumed to be valid for *all* values of m and n. Therefore,

$$a^0 = \frac{a^n}{a^n} = 1 \tag{1.6}$$

$$a^{-n} = \frac{a^0}{a^n} = \frac{1}{a^n} \tag{1.7a}$$

$$a^n = \frac{1}{a^{-n}} \tag{1.7b}$$

As an illustration, we have the following:

$$\frac{a^{2.5}(a^{1.2})^3}{a^{0.7}} = \frac{a^{2.5}a^{3.6}}{a^{0.7}} = a^{2.5}a^{3.6}a^{-0.7} = a^{5.4}$$

From Eq. (1.2), we have

$$(a^{1/n})^n = (a^n)^{1/n} = a^1 = a$$

If $a = b^n$, then $a^{1/n} = b$, and we say that b is the *nth root of a*, with these exceptions: If $n = 2$, b is the *square root* of a; if $n = 3$, b is the *cube root* of a. Alternatively, the nth root of a can be denoted by $\sqrt[n]{a}$. In this notation, the number a is called the *radicand* and n is called the *index*. Thus, $\sqrt[n]{a} = a^{1/n}$. We also have

$$a^{m/n} = (a^{1/n})^m = (a^m)^{1/n}$$

Consider the quantity e^A, where e is the number defined in Art. 1.20. If the expression for A is rather involved or contains exponents, it is convenient to replace the symbol e^A with the expression "exp A." For example,

$$8 + 3 \exp x^2 = 8 + 3e^{x^2}$$

If a and n are real numbers, the value of a^n can be obtained by means of the calculator.

EXAMPLE 1.5 Find the value of a (to five significant figures) in each of the following cases: $a^5 = 47$; $a^{3.6} = 52$; $a^{-3.2} = 0.007$.

SOLUTION

Case 1: $a^5 = 47$ $\quad a = 47^{1/5} = 47^{0.20} = 2.1598$

Case 2: $a^{3.6} = 52$ $\quad a = 52^{1/3.6} = 2.9969$

Case 3: $a^{-3.2} = 0.007$ $\quad a^{3.2} = 1/0.007 = 142.857$

$$a = 142.857^{1/3.2} = 4.7142$$

These results can be verified instantly, for we have

$$2.1598^5 = 47 \qquad 2.9969^{3.6} = 52 \qquad 4.7142^{-3.2} = 0.007$$

The quantity $(-1)^n$ has the value $+1$ if n is an even integer and -1 if n is an odd integer.

1.8 LOGARITHMS

Let b denote a positive number other than 1 and let $b^c = x$. Then c is the *logarithm of* x to the base b. Expressed symbolically, $c = \log_b x$. Thus, the logarithm of a number x is the *power* to which the base b must be raised to obtain x. For example, since $6^3 = 216$, then $\log_6 216 = 3$. Because b is positive, x is also positive, and the definition of a logarithm applies solely to positive numbers.

The following laws of logarithms stem from the laws of exponents:

$$\log_b xy = \log_b x + \log_b y \tag{1.8}$$

$$\log_b \frac{x}{y} = \log_b x - \log_b y \tag{1.9}$$

$$\log_b x^u = u \log_b x \tag{1.10}$$

$$\log_b x = \frac{1}{\log_x b} \tag{1.11}$$

From Eq. (1.9) we deduce the following relationships, which are independent of the base: $\log x > 0$ if $x > 1$; $\log x = 0$ if $x = 1$; $\log x < 0$ if $x < 1$.

From the definition of a logarithm, we have $\log_b b = 1$ and $\log_b b^m = m$.

Two systems of logarithms are widely used. The *common* or *Briggs system* uses 10 as base; the *natural* or *napierian system* uses as base the number e, which is defined in Art. 1.20. The notation $\ln x$ means $\log_e x$. Both common and natural logarithms are directly obtainable by use of the calculator. If the logarithm of a number in some other system is required, the value can be obtained by applying the following equation:

$$\log_b x = \frac{\log_a x}{\log_a b} \tag{1.12}$$

This relationship provides a means of converting the logarithm of a number from one system to another.

EXAMPLE 1.6 Evaluate $\log_3 17$ and verify the result.

SOLUTION We shall solve this problem without applying Eq. (1.12) explicitly in order to acquire greater dexterity in working with logarithms. Let $c = \log_3 17$. From the definition of a logarithm, we have

$$3^c = 17$$

Equating the logarithms of the two sides of the equation and then applying Eq. (1.10), we obtain

$$c \log_{10} 3 = \log_{10} 17$$

$$c = \frac{\log_{10} 17}{\log_{10} 3} = \frac{1.230449}{0.477121} = 2.57890$$

Alternatively, the value of c can be obtained by applying Eq. (1.12).
As proof of this result, we obtain the following by calculator: $3^{2.57890} = 17$.

EXAMPLE 1.7 If $\log_a 8 = 2.2694$, what is a?

SOLUTION Applying the definition of a logarithm, we have

$$a^{2.2694} = 8 \qquad a = 8^{1/2.2694} = 2.5$$

EXAMPLE 1.8 If $6^x = 1{,}400{,}000$, what is x?

SOLUTION In effect, the given information states that

$$x = \log_6 1{,}400{,}000$$

Thus, this example is similar to Example 1.6. Proceeding as before, we have

$$x \log 6 = \log 1{,}400{,}000$$

$$x = 7.8984$$

EXAMPLE 1.9 As time elapses, a quantity x increases in magnitude in this manner:

$$x = 120e^{0.08t}$$

where e is the base of natural logarithms and t denotes elapsed time in days. How long will it take for x to attain the value of 300? Verify the solution.

SOLUTION Rearranging the given equation and then setting $x = 300$, we obtain

$$e^{0.08t} = \frac{x}{120} = \frac{300}{120} = 2.5$$

From the definition of the natural logarithm, we have

$$0.08t = \ln 2.5 = 0.9163$$

Solving, $$t = 11.45 \text{ days}$$

The solution is verified by substituting this value of t in the given equation, and the result is $x = 300$.

1.9 OPERATIONS WITH IMAGINARY AND COMPLEX NUMBERS

Let a denote a number that is real and positive. By Eq. (1.3), we have the following:

$$\sqrt{-a} = \sqrt{(-1)a} = \sqrt{-1}\sqrt{a}$$

The number $\sqrt{-1}$ is taken as the unit of imaginary numbers, and it is denoted by i. (In electrical engineering, the designation j is used.) Then $i^2 = -1$, and

$$\sqrt{-a} = i\sqrt{a}$$

Starting with i^0 and multiplying successively by i, we obtain the following:

$$i^0 = 1 \qquad i^1 = i \qquad i^2 = -1 \qquad i^3 = -i$$

This set of values then repeats itself cyclically. Therefore, if n is any integer, positive or negative, we have

$$i^{4n} = i^0 = 1 \qquad i^{4n+1} = i^1 = i$$

$$i^{4n+2} = i^2 = -1 \qquad i^{4n+3} = i^3 = -i$$

EXAMPLE 1.10 Evaluate each of the following quantities: i^{12}, i^{19}, i^{25}, i^{-17}.

SOLUTION

$$12 = 4 \times 3 \qquad \therefore \quad i^{12} = i^0 = 1$$

$$19 = 4 \times 4 + 3 \qquad \therefore \quad i^{19} = i^3 = -i$$

$$25 = 4 \times 6 + 1 \qquad \therefore \quad i^{25} = i^1 = i$$

$$-17 = 4(-5) + 3 \qquad \therefore \quad i^{-17} = i^3 = -i$$

In the following material, all letters except i denote real numbers. A complex number, which is defined in Art. 1.1, can be represented by $a + ib$. We have the following:

$$(a + ib) + (c + id) = (a + c) + i(b + d)$$

$$(a + ib) - (c + id) = (a - c) + i(b - d)$$

$$(a + ib)(c + id) = (ac - bd) + i(ad + bc)$$

The numbers $a + ib$ and $a - ib$ are called *conjugates* of each other. Then

$$(a + ib) + (a - ib) = 2a$$

$$(a + ib) - (a - ib) = 2ib$$

$$(a + ib)(a - ib) = a^2 + b^2$$

Assume that a fraction has a complex number as its denominator. If we multiply the numerator and denominator of this fraction by the conjugate of the denominator, we obtain a fraction that is equal to the given one and that has a real number as its denominator.

EXAMPLE 1.11 The following fraction is given:

$$\frac{7 - i4}{6 - i9}$$

Transform this fraction to one that has the same value but contains a real denominator. Verify the result.

SOLUTION

$$\frac{7-i4}{6-i9}=\frac{(7-i4)(6+i9)}{(6-i9)(6+i9)}=\frac{(42+36)+i(63-24)}{6^2+9^2}=\frac{78+i39}{117}=\frac{2+i}{3}$$

Verification. That the final fraction is equal to the original one can be proved by cross-multiplying the two fractions. (Refer to Art. 1.6.) The cross-multiplication yields these values:

$$(7-i4)3=21-i12$$

$$(6-i9)(2+i)=21-i12$$

The result is thus confirmed.

In general,

$$\frac{a+ib}{c+id}=\frac{(ac+bd)+i(bc-ad)}{c^2+d^2}$$

The multiplication, division, and exponentiation of complex numbers can also be performed by expressing these numbers in polar form, in the manner demonstrated in Arts. 6.2.12 and 6.2.13.

1.10 PROPORTIONS AND PROPORTIONALITIES

The notation $a:b=c:d$ is read "a is to b as c is to d," and it means that $a/b=c/d$. Thus, a and b have the same proportion (or ratio) as c and d.

The notation $y \propto x$ means that y is directly proportional to x, or y varies directly as x. The relationship can be expressed as $y=kx$, where k is a constant. Similarly, the notation $y \propto 1/x$ means that y is inversely proportional to x, or y varies inversely as x. The relationship can be expressed as $y=k/x$.

Numerical problems pertaining to variation can be solved by expressing the given information in the form of an equation by use of the constant k.

EXAMPLE 1.12 A simply supported steel shaft of diameter D and length L is subjected to a uniformly distributed load W. The deflection d of the shaft varies directly as W, directly as L^3, and inversely as D^4. The deflection is 3.88 cm when $D=10$ cm, $L=6$ m, and $W=14$ kN. Find the deflection when $D=11.5$ cm, $L=7$ m, and $W=13.2$ kN.

SOLUTION The equation for d becomes

$$d=\frac{kWL^3}{D^4}$$

Applying the subscripts 1 and 2 to denote the first and second set of conditions, respectively, we have

$$\frac{d_2}{d_1} = \frac{W_2}{W_1}\left(\frac{L_2}{L_1}\right)^3\left(\frac{D_1}{D_2}\right)^4$$

$$d_2 = 3.88\left(\frac{13.2}{14}\right)\left(\frac{7}{6}\right)^3\left(\frac{10}{11.5}\right)^4 = 3.32 \text{ cm}$$

1.11 DEFINITIONS PERTAINING TO EQUATIONS

An equation that contains letters representing numbers is described as *literal.* A literal equation is termed an *identity* if arbitrary values can be assigned to the letters, and a *conditional equation* if such is not the case. For example, the equations $(x + y)(x - y) = x^2 - y^2$ and $\sin^2 x + \cos^2 x = 1$ are universally true, and therefore they are identities. On the other hand, the equation $x^2 + 2x = 15$ is valid only if x is 3 or -5, and therefore it is a conditional equation. Some authors restrict the symbol $=$ to a conditional equation and use the symbol $\equiv$ for an identity.

The letters in a conditional equation are called *unknowns,* and a set of values of the unknowns that satisfies the equation is termed a *solution.* Thus, $x = 3$, $y = -7$ is a solution of the equation $x^2 + 2y + 5 = 0$. If the conditional equation contains a single unknown, the solution is termed a *root* of the equation. For example, 11 and -2 are roots of the equation $x^2 - 9x - 22 = 0$.

A *polynomial equation* is formed when a polynomial is set equal to 0. The degree of a polynomial equation equals that of the polynomial itself. For example, the equation $x^2y + 5x - 7y = 0$ is of the third degree. Equations of the first, second, third, and fourth degrees are known, respectively, as *linear, quadratic, cubic,* and *quartic* (or *biquadratic*) equations.

1.12 SIMULTANEOUS LINEAR EQUATIONS

A system containing n simultaneous linear equations with n unknowns can be solved by applying certain general procedures. However, if n is 2 or 3, the solution can be obtained directly by means of the calculator. The reader should become fully familiar with the procedure.

EXAMPLE 1.13 Solve each of the following systems of simultaneous equations:

System 1

$$3x + 4y = 19$$

$$7x - 6y = -63$$

System 2

$$5x - 3y + 6z = -91$$

$$2x + 8y - 7z = 84$$

$$4x - 2y - 5z = -30$$

SOLUTION The solution to system 1 is $x = -3$, $y = 7$, and the solution to system 2 is $x = -8$, $y = 9$, $z = -4$. These results can be tested by substituting the numerical values in the given equations.

1.13 NOTATION FOR FUNCTIONS, SUMS, AND PRODUCTS

The notation $f(x)$ denotes a function of x, and $f(r)$ denotes the value of $f(x)$ when $x = r$.

EXAMPLE 1.14 If $f(x) = 4x^3 + 7x^2 + 9$, what is $f(5)$?

SOLUTION Replacing x with 5, we obtain

$$f(5) = 4(5^3) - 7(5^2) + 9 = 334$$

Where several functions of x are present, they can be assigned the individual designations $f(x)$, $g(x)$, $h(x)$, etc.

The Greek letter Σ is used to denote the sum of a group of expressions of identical structure in which the variable increases successively by 1 as we proceed from one expression to the following one. Entries placed above and below Σ identify the variable and indicate its range of values.

EXAMPLE 1.15 Evaluate the following expression (to four significant figures):

$$\sum_{x=3}^{7} \frac{x^2}{x+1}$$

SOLUTION Let S denote the value of this expression. The variable x ranges from 3 to 7, inclusive, and we have

$$S = \frac{3^2}{4} + \frac{4^2}{5} + \frac{5^2}{6} + \frac{6^2}{7} + \frac{7^2}{8} = 20.88$$

The following laws apply:

$$\sum [f(x) + g(y)] = \sum f(x) + \sum g(y)$$

$$\sum af(x) = a \sum f(x)$$

where a denotes a constant.

Where two variables are multiplied, the letter Σ is used twice. As an illustration, we have

$$\sum_{x=4}^{7} \sum_{y=1}^{4} x^2 y = 4^2 \cdot 1 + 5^2 \cdot 2 + 6^2 \cdot 3 + 7^2 \cdot 4 = 370$$

Similarly, the Greek letter Π is used to denote the product of a group of expressions of identical structure in which the variable increases successively by 1. For example,

$$\prod_{u=2}^{5} (x + u) = (x + 2)(x + 3)(x + 4)(x + 5)$$

1.14 FACTORIAL NUMBERS

Let n denote a positive integer. The symbol $n!$ (read "n factorial") denotes the product of the first n integers. The integers are usually recorded in descending order of magnitude. For example,

$$5! = 5 \cdot 4 \cdot 3 \cdot 2 \cdot 1 = 120$$

The product of a group of consecutive integers can be expressed as the quotient of two factorial numbers. For example,

$$17 \cdot 16 \cdot 15 \cdot 14 \cdot 13 = \frac{17!}{12!}$$

A double-factorial notation is used to denote the product of even or odd integers exclusively, in this manner:

$$(2n)!! = (2n)(2n - 2)(2n - 4) \cdots 6 \cdot 4 \cdot 2$$
$$(2n - 1)!! = (2n - 1)(2n - 3)(2n - 5) \cdots 5 \cdot 3 \cdot 1$$

EXAMPLE 1.16 Evaluate 10!! and 9!!.

SOLUTION

$$10!! = 10 \cdot 8 \cdot 6 \cdot 4 \cdot 2 = 3840$$
$$9!! = 9 \cdot 7 \cdot 5 \cdot 3 \cdot 1 = 945$$

1.15 PROGRESSIONS AND SERIES

A sequence of numbers (or expressions that represent numbers) that is generated in a prescribed manner is termed a *progression.* When these numbers are connected with plus or minus signs, they constitute a *series.* We shall let u_k denote the kth term in the series, n the number of terms, and S the algebraic sum of the series.

A series is described as *finite* or *infinite* according to whether n is finite or infinite, respectively. An infinite series is *convergent* if S approaches a limiting value as n increases beyond bound, and *divergent* if such is not the case. An *alternating series* is one in which plus and minus signs alternate.

There is a type of problem in which we are given several terms in the series and we must establish the value of some subsequent term or find the sum of the series. To solve a problem of this type, we must compare successive terms to recognize the pattern that is inherent in the series. In the following material, all letters except x denote constants.

EXAMPLE 1.17 Setting $n = 29$, find the sum of the following alternating series:

$$4 - 7 + 10 - 13 + 16 - 19 + \cdots$$

SOLUTION The basic property of this series is that the terms increase in absolute value by 3. Therefore, the kth term is

$$u_k = (-1)^{k-1}[4 + 3(k - 1)] = (-1)^{k-1}(3k + 1)$$

We can pair consecutive terms, and the result is as follows:

$$u_1 + u_2 = u_3 + u_4 = \cdots = u_{27} + u_{28} = -3$$

The number of pairs is $28/2 = 14$. The 29th term is $+(3 \times 29 + 1) = +88$. Then

$$S = 14(-3) + 88 = 46$$

Alternatively, we can pair u_2 and u_3, u_4 and u_5, etc.; each pair has a sum of 3. Then

$$S = u_1 + 14 \times 3 = 4 + 14 \times 3 = 46$$

This alternative method of solution is the simpler of the two because it obviates the need to find u_k.

A *power series* is one having the form

$$A_0 + A_1(x - c) + A_2(x - c)^2 + \cdots + A_n(x - c)^n$$

However, some coefficients may be 0.

The following types of series arise very frequently:

Arithmetic Series. This is one in which $u_{k+1} - u_k$ is constant. Let d denote this constant. The general form of the series is

$$a + (a + d) + (a + 2d) + \cdots + [a + (n - 1)d]$$

The kth term is $a + (k - 1)d$, and the sum is

$$S = \frac{n}{2}(u_1 + u_n) = \frac{n}{2}[2a + (n - 1)d] \tag{1.13}$$

EXAMPLE 1.18 The following series is given:

$$-85 - 81 - 77 - 73 - \cdots$$

Find the thirteenth term of this series, and find the sum of the first 18 terms.

SOLUTION This is an arithmetic series in which $a = -85$ and $d = 4$. Then

$$u_{13} = -85 + 12 \times 4 = -37$$

With $n = 18$, we have

$$S = \frac{18}{2}[2(-85) + 17 \times 4] = -918$$

EXAMPLE 1.19 Find the sum of the first 19 terms of this series:

$$27 + 22 + 17 + 12 + \cdots$$

SOLUTION This is an arithmetic series in which $a = 27$ and $d = -5$. With $n = 19$,

$$S = \frac{19}{2}[2 \times 27 + 18(-5)] = -342$$

EXAMPLE 1.20 An arithmetic series has -3 as its first term and 37 as its last term. The sum of the series is 102. What is the difference between successive terms?

SOLUTION By Eq. (1.13),

$$S = \frac{n}{2}(-3 + 37) = 102 \qquad n = 6$$

$$u_6 = u_1 + (n-1)d \qquad \text{or} \qquad 37 = -3 + 5d$$

Then $d = 8$.

The series is

$$-3 + 5 + 13 + 21 + 29 + 37$$

Geometric Series. This is one in which the ratio u_{k+1}/u_k is constant. Let r denote this constant. The general form of the series is

$$a + ar + ar^2 + \cdots + ar^{n-1}$$

The kth term is ar^{k-1}, and the sum is

$$S = \frac{a(r^n - 1)}{r - 1} \tag{1.14}$$

This equation reveals that an infinite geometric series is convergent if and only if $|r| < 1$, and the limit of the sum is

$$S = \frac{a}{1 - r} \tag{1.14a}$$

EXAMPLE 1.21 Find the sum of the first 10 terms of this series:

$$0.2 + 0.8 + 3.2 + 12.8 + \cdots$$

SOLUTION This is a geometric series in which $a = 0.2$ and $r = 4$. With $n = 10$,

$$S = \frac{(0.2)(4^{10} - 1)}{3} = 69{,}905$$

EXAMPLE 1.22 Find the limiting value of S for the following series:

$$8 + \frac{8}{3} + \frac{8}{3^2} + \frac{8}{3^3} + \frac{8}{3^4} + \cdots$$

SOLUTION This is a geometric series in which $a = 8$ and $r = 1/3$. The series is convergent because $|r| < 1$, and Eq. (1.14*a*) yields

$$S = \frac{8}{1 - 1/3} = \frac{8}{2/3} = 12$$

EXAMPLE 1.23 Find the limiting value of S for the following series:

$$8 - \frac{8}{3} + \frac{8}{3^2} - \frac{8}{3^3} + \frac{8}{3^4} - \cdots$$

SOLUTION This is a geometric series in which $a = 8$ and $r = -1/3$. Equation (1.14*a*) yields

$$S = \frac{8}{1 - (-1/3)} = \frac{8}{4/3} = 6$$

EXAMPLE 1.24 A geometric series has these characteristics: The first term is 5, the ratio of one term to the preceding term is 4, and the sum of the first n terms is 109,225. What is n?

SOLUTION Substituting in Eq. (1.14), we obtain

$$\frac{5(4^n - 1)}{4 - 1} = 109{,}225$$

$$4^n = \frac{109{,}225 \times 3}{5} + 1 = 65{,}536$$

$$n \log 4 = \log 65{,}536 \qquad n = 8$$

Harmonic Series. This is one in which the reciprocals of the terms form an arithmetic progression. The following series is illustrative:

$$\frac{1}{3} + \frac{1}{5} + \frac{1}{7} + \cdots + \frac{1}{2n + 1}$$

Arithmogeometric Series. This is a composite of an arithmetic and geometric series, the general form being

$$a + (a + d)r + (a + 2d)r^2 + \cdots + [a + (n - 1)d]r^{n-1}$$

The sum is

$$S = \frac{a(r^n - 1)}{r - 1} + \frac{dr}{(r - 1)^2}\left[(n - 1)r^n - nr^{n-1} + 1\right] \tag{1.15}$$

This equation reveals that an infinite arithmogeometric series is convergent if $|r| < 1$, and the limit of the sum is

$$S = \frac{a}{1 - r} + \frac{dr}{(1 - r)^2} \tag{1.15a}$$

EXAMPLE 1.25 Find the limiting value of S for the following series:

$$25 - 27(0.8) + 29(0.8)^2 - 31(0.08)^3 + \cdots$$

SOLUTION This is an arithmogeometric series in which $a = 25$, $d = 2$, and $r = -0.8$. The series is convergent because $|r| < 1$, and Eq. (1.15*a*) yields the limiting value

$$\frac{25}{1.8} + \frac{2(-0.8)}{(1.8)^2} = 13.3951$$

Polynomial Progression. This is one in which u_k is a polynomial function of k, by an equation of this form:

$$u_k = a_0 k^m + a_1 k^{m-1} + a_2 k^{m-2} + \cdots + a_m$$

where m is a positive integer, the a's are all constant, and $a_0 \neq 0$. The progression is said to be of the mth degree.

The difference between two successive terms in a polynomial progression is called a *first difference*; the difference between two successive first differences is called a *second difference*; etc. Each set of differences is itself a polynomial progression, and the degree of the progression diminishes by 1 whenever we form new differences. The following principle emerges:

Theorem 1.1. If a polynomial progression is of the mth degree, the mth differences are all equal. Conversely, if the mth differences are all equal, the polynomial progression is of the mth degree.

EXAMPLE 1.26 The following are the first seven terms of a polynomial progression: -4, 3, 26, 77, 168, 311, 518. Establish the degree of the progression, and find the eighth term.

SOLUTION In Table 1.1, we have recorded the given terms on the first line, the differences between these terms on the second line, the second differences on the third line, etc. For example, the values on the second line are obtained by these calculations: $3 - (-4) = 7$, $26 - 3 = 23$, etc. The values on the third line are obtained by these calculations: $23 - 7 = 16$, $51 - 23 = 28$, etc. Finally, the values on the fourth line are obtained by these calculations: $28 - 16 = 12$, $40 - 28 = 12$, etc. The differences are recorded midway between the corresponding numbers in the line above. We find that the third differences are all equal, and it follows that the progression is of the third degree. To find u_8, we proceed in this manner:

$$a_5 = 12 \qquad b_6 = 64 + 12 = 76 \qquad c_7 = 207 + 76 = 283$$

$$u_8 = 518 + 283 = 801$$

TABLE 1.1 Finding the Degree of a Polynomial Progression

−4		3		26		77		168		311		518		u_8
	7		23		51		91		143		207		c_7	
		16		28		40		52		64		b_6		
			12		12		12		12		a_5			

Although we shall not undertake this task, it is possible to formulate the polynomial expression on the basis of the successive differences. In Example 1.26, the relationship is

$$u_k = 2k^3 - 4k^2 + 5k - 7$$

Fibonacci Series. This "series" is a progression that is generated by this formula:

$$u_1 = u_2 = 1 \qquad \text{and} \qquad u_k = u_{k-1} + u_{k-2}$$

Expressed verbally, each term beyond the second is the sum of the two preceding terms. The first 12 terms of the series are as follows: 1, 1, 2, 3, 5, 8, 13, 21, 34, 55, 89, 144. The Fibonacci series has numerous practical applications.

Three consecutive terms in this series are also related in this manner:

$$u_k^2 = u_{k+1}u_{k-1} + (-1)^{k-1}$$

It follows from this relationship that the ratio u_k/u_{k-1} approaches a limiting value as k increases beyond bound. This limiting value is $(1 + \sqrt{5})/2 = 1.618034$ to six decimal places. The ratio approaches this limit rapidly. For example, setting $k = 12$, we have $u_{12}/u_{11} = 144/89 = 1.617978$ to six decimal places. Thus, the Fibonacci series approaches a geometric progression as the number of terms increases.

1.16 BINOMIAL COEFFICIENTS

Let n and r denote nonnegative integers, with $n \geq r$. For a reason that will soon become apparent, the expression $n!/[r!(n-r)!]$ is termed a *binomial coefficient*, and it is denoted by this notation:

$$\frac{n!}{r!(n-r)!} = \binom{n}{r}$$

The expression at the right is read "nCr." It follows at once that

$$\binom{n}{r} = \binom{n}{n-r} \tag{1.16}$$

EXAMPLE 1.27 Evaluate the following binomial coefficient:

$$\binom{12}{7}$$

SOLUTION

$$\binom{12}{7} = \frac{12!}{7!\,5!} = \frac{12 \cdot 11 \cdot 10 \cdot 9 \cdot 8}{5 \cdot 4 \cdot 3 \cdot 2} = 792$$

1.17 BINOMIAL THEOREM

Let n denote a positive integer. The expansion of $(a + b)^n$ yields the following series:

$$(a+b)^n = a^n + \frac{n}{1!}a^{n-1}b + \frac{n(n-1)}{2!}a^{n-2}b^2 + \frac{n(n-1)(n-2)}{3!}a^{n-3}b^3 + \cdots$$
$$+ \frac{n(n-1)(n-2)\cdots(n-k+1)}{k!}a^{n-k}b^k + \cdots + b^n \qquad (1.17)$$

Thus, the series contains $n + 1$ terms. In the kth term beyond the first, the coefficient can be written as $n!/[k!(n-k)!]$. Therefore, the expansion of $(a + b)^n$ can also be expressed in terms of the binomial coefficients defined in Art. 1.16, and the series becomes

$$(a+b)^n = \binom{n}{0}a^n + \binom{n}{1}a^{n-1}b + \binom{n}{2}a^{n-2}b^2$$
$$+ \binom{n}{3}a^{n-3}b^3 + \cdots + \binom{n}{k}a^{n-k}b^k + \cdots + \binom{n}{n}b^n \qquad (1.17a)$$

Equation (1.16) reveals that the coefficients in this equation are arranged symmetrically about the center.

EXAMPLE 1.28 Write the expansion of $(c + x)^5$.

SOLUTION The series will contain 6 terms. Therefore, the number of coefficients to be evaluated is $6/2 - 1 = 2$. They are as follows:

$$\frac{5}{1!} = 5 \qquad \frac{5 \cdot 4}{2!} = 10$$

Replacing a with c and b with x in Eq. (1.17), we obtain

$$(c+x)^5 = c^5 + 5c^4x + 10c^3x^2 + 10c^2x^3 + 5cx^4 + x^5$$

The coefficients in the binomial expansion can be obtained quickly by constructing *Pascal's triangle*, which is shown in Table 1.2 for values through $n = 7$. The

TABLE 1.2 Pascal's Triangle

n	Binomial coefficients
0	1
1	1 1
2	1 2 1
3	1 3 3 1
4	1 4 6 4 1
5	1 5 10 10 5 1
6	1 6 15 20 15 6 1
7	1 7 21 35 35 21 7 1

coefficients are generated by this formula: Let $c_{n,r}$ denote the rth coefficient when the exponent is n. Then

$$c_{n,r} + c_{n,r+1} = c_{n+1,r+1}$$

For example,

$$c_{6,5} + c_{6,6} = 15 + 6 = 21 = c_{7,6}$$

Equation (1.17) is also valid if n is nonintegral or negative and $|b/a| < 1$. However, in this case the binomial expansion is an infinite series, and the term b^n never arises.

1.18 POLYNOMIAL EQUATIONS IN A SINGLE UNKNOWN

A polynomial equation of nth degree in a single unknown x can be written in the following manner, which is known as the *p-form* of the equation:

$$x^n + p_1 x^{n-1} + p_2 x^{n-2} + \cdots + p_n = 0$$

This equation is described as real, rational, and integral if all the p's are real, rational, and integral.

The polynomial equation of nth degree has n roots, but some may be identical. Let $r_1, r_2, r_3, \ldots, r_n$ denote the roots. The following principles apply:

Theorem 1.2. If r_i is a root of the equation $f(x) = 0$, then $x - r_i$ is a factor of $f(x)$.

The converse of this statement is also valid, and therefore the polynomial equation can be recast in this *factored form*:

$$(x - r_1)(x - r_2)(x - r_3) \cdots (x - r_n) = 0 \tag{1.18}$$

EXAMPLE 1.29 Construct an equation that has 2, -3, and 7 as its roots.

SOLUTION In factored form, the equation is

$$(x - 2)(x + 3)(x - 7) = 0$$

Performing the multiplication, we obtain the following p-form:

$$x^3 - 6x^2 - 13x + 42 = 0$$

This result can be verified by substituting each given value for x.

Theorem 1.3. If a real, rational, integral equation has a root r_i that is real and rational, then r_i is an integer and a factor of p_n.

Theorem 1.4. If the complex number $a + ib$ is the root of a real equation, the conjugate complex number $a - ib$ is also a root.

EXAMPLE 1.30 Construct a real equation that has 4 and $-2 - i7$ as roots.

SOLUTION To be real, the equation must also have $-2 + i7$ as a root. Therefore, the factored form of the equation is

$$(x - 4)(x + 2 + i7)(x + 2 - i7) = 0$$

The product of the second and third terms is

$$(x + 2)^2 + 7^2 = x^2 + 4x + 53$$

Multiplying this trinomial by $x - 4$, we obtain the following *p*-form:

$$x^3 + 37x - 212 = 0$$

Theorem 1.5. If an irrational number of the form $a + \sqrt{b}$ is a root of a rational equation, the irrational number $a - \sqrt{b}$ is also a root.

EXAMPLE 1.31 Construct a rational equation that has -8 and $3 + 2\sqrt{7}$ as roots.

SOLUTION To be rational, the equation must also have $3 - 2\sqrt{7}$ as a root. The factored form of the equation is

$$(x + 8)(x - 3 - 2\sqrt{7})(x - 3 + 2\sqrt{7}) = 0$$

The product of the second and third terms is

$$(x - 3)^2 - 4 \times 7 = x^2 - 6x - 19$$

Multiplying this trinomial by $x + 8$, we obtain the following *p*-form:

$$x^3 + 2x^2 - 67x - 152 = 0$$

1.19 SOLUTION OF QUADRATIC EQUATION

A second-degree equation in a single unknown x has the general form

$$ax^2 + bx + c = 0$$

where a, b, and c are constants and $a \neq 0$. This equation has two roots, which we denote by x_1 and x_2. Let

$$f = b^2 - 4ac$$

The roots of the equation are as follows:

$$x_1 = \frac{-b + \sqrt{f}}{2a} \qquad x_2 = \frac{-b - \sqrt{f}}{2a}$$

EXAMPLE 1.32 Solve the equation $8x^2 + 10x - 63 = 0$.

SOLUTION

$$a = 8 \qquad b = 10 \qquad c = -63$$

$$f = 100 - 4(8)(-63) = 2116 \qquad \sqrt{f} = 46$$

$$x_1 = \frac{-10 + 46}{16} = 2.25 \qquad x_2 = \frac{-10 - 46}{16} = -3.5$$

EXAMPLE 1.33 Solve the equation $x^2 - x\sqrt{3} - 6 = 0$.

SOLUTION

$$a = 1 \qquad b = -\sqrt{3} \qquad c = -6$$

$$f = 3 - 4(1)(-6) = 27 \qquad \sqrt{f} = 3\sqrt{3}$$

The roots are $2\sqrt{3}$ and $-\sqrt{3}$.

Assume that a, b, and c are real numbers. The character of the two roots is determined by the parameter f, and f is accordingly termed the *discriminant* of the equation. If $f > 0$, the two roots are real and unequal. If $f = 0$, the two roots are real and equal. If $f < 0$, the two roots are conjugate complex numbers.

Where the roots of a quadratic equation are real, they can be obtained by calculator.

1.20 DEFINITION OF e

The quantity e is defined in this manner:

$$e = \lim_{n \to \infty} \left(1 + \frac{1}{n}\right)^n$$

By expanding the binomial in accordance with Eq. (1.17) and allowing n to become infinite, we obtain the following infinite series:

$$e = 1 + \frac{1}{1!} + \frac{1}{2!} + \frac{1}{3!} + \frac{1}{4!} + \cdots \tag{1.19}$$

The quantity e is an irrational number. To six significant figures, its value is 2.71828. Similarly, we have

$$e^x = \lim_{n \to \infty} \left(1 + \frac{1}{n}\right)^{nx}$$

Again expanding and allowing n to become infinite, we obtain

$$e^x = 1 + \frac{x}{1!} + \frac{x^2}{2!} + \frac{x^3}{3!} + \frac{x^4}{4!} + \cdots \tag{1.20}$$

If x is a real number, the value of e^x can be obtained by means of the calculator.

1.21 FLOATING-DECIMAL-POINT SYSTEM

In the Arabic numeral system, the position occupied by a digit corresponds to a specific power of 10. For example,

$$4725.9 = 4(1000) + 7(100) + 2(10) + 5(1) + 9\left(\frac{1}{10}\right)$$

$$= 4(10^3) + 7(10^2) + 2(10^1) + 5(10^0) + 9(10)^{-1}$$

In expressing a number, the decimal point can be moved m places to the right if we then multiply by 10^{-m}, and it can be moved n places to the left if we then multiply by 10^n. For example,

$$0.0572 = 572(10^{-4}) \qquad 873.1 = 8.731(10^2)$$

Where numbers of widely varying magnitude are present, it may be desirable to express the numbers in a standard form in which the decimal point is placed at some assigned position in the sequence of digits. This procedure can be followed if each number is then multiplied by the appropriate power of 10. Moreover, for simplicity, the notation (10^n) is replaced with En. Therefore, we may regard En as an instruction to move the decimal point in the preceding number n places *to the right* to obtain the number in natural form. For example,

$$389.16 = 0.38916\text{E}3 = 3891.6\text{E–}1$$

The foregoing method of expressing numbers is known as the *floating-decimal-point system.*

EXAMPLE 1.34 Record the following numbers in the floating-decimal-point system with the decimal point following the first digit: 47,281, 0.00296, 3.972.

SOLUTION The numbers are 4.7281E4, 2.96E–3, and 3.972E0.

Where numbers are to be added or subtracted under the floating-decimal-point system, the exponents of 10 must first be made identical.

EXAMPLE 1.35 Perform the following operation:

$$29.7\text{E–}2 + 3.1356\text{E}1 + 0.148302\text{E}3$$

Express the sum in each of the following forms: first, with the decimal point immediately following the first digit and second, in natural form.

SOLUTION Let S denote the sum. We shall select the smallest exponent, which is -2, as the standard. Then

$$S = 29.7\text{E–}2 + 3135.6\text{E–}2 + 14{,}830.2\text{E–}2$$

$$= (29.7 + 3135.6 + 14{,}830.2)\text{E–}2 = 17{,}995.5\text{E–}2$$

In the specified forms, $S = 1.7955\text{E}2 = 179.955$.

1.22 NONDECIMAL NUMERAL SYSTEMS

The decimal numeral system uses 10 digits (0 to 9, inclusive), but it is possible to devise a numeral system that uses any feasible number of digits. Specifically, the binary system uses 2 digits (0 and 1), and the octal system uses 8 digits (0 to 7, inclusive). The hexadecimal system uses 16 digits, and the digits beyond 9 are represented by uppercase letters, in this manner: A for 10, B for 11, C for 12, D for 13, E for 14, and F for 15.

The number of digits used in a numeral system is called the *base* (or *radix*) of the system, and the base is displayed by means of a subscript appended to the number. For example, 5204_8 is a number in the octal system. Where the subscript is omitted, it is understood to be 10. We shall denote the base of a system by B.

In a *positional* numeral system, a number is expressed as the sum of powers of B, and the digits are so arranged that the corresponding powers of B diminish consecutively from left to right. If the number is nonintegral, the digits corresponding to the powers 0 and -1 are separated by a period, which is termed a *decimal point* in the decimal system.

It is frequently necessary to convert a number from one numeral system to another. The conversion from decimal to binary, octal, or hexadecimal, or vice versa, can be obtained by calculator. We shall investigate the conversion to some other system, but we shall confine our investigation to integral numbers for brevity.

EXAMPLE 1.36 Express 46025_7 as a decimal number.

SOLUTION This form of conversion is straightforward. Let N denote the number. Taking the digits in reverse order, we have

$$N = 5 + 2(7) + 0(7^2) + 6(7^3) + 4(7^4) = 11681_{10}$$

We now consider the reverse form of conversion. Let N denote a positive integral nondecimal number, and assume it has the form $(d_2 d_1 d_0)_B$. Taking the digits in reverse order, we have

$$N = d_0 + d_1 B + d_2 B^2$$

If we divide N by B, we obtain a quotient $Q_1 = d_1 + d_2 B$ and a remainder d_0. If we now divide Q_1 by B, we obtain a quotient $Q_2 = d_2$ and a remainder d_1. If we now divide Q_2 by B, we obtain a quotient 0 and a remainder d_2. Thus, this iterative form of division by B yields the required digits in reverse order.

TABLE 1.3 Conversion of a Number

Division	Quotient	Remainder
56681/7	8097	2
8097/7	1156	5
1156/7	165	1
165/7	23	4
23/7	3	2
3/7	0	3

EXAMPLE 1.37 Express 56681_{10} as a number in the numeral system having 7 as base.

SOLUTION The successive divisions are performed in Table 1.3. Taking the digits in reverse order, we have

$$56681_{10} = 324152_7$$

This result is easily verified by converting it to decimal form.

1.23 PERMUTATIONS

The study of permutations rests on the following principle, which is known as the *law of multiplication.*

Theorem 1.6. Assume that n acts are to be performed in sequence. If the first act can be performed in m_1 alternative ways, the second act in m_2 alternative ways, ..., the nth act in m_n alternative ways, the entire set of n acts can be performed in $m_1 m_2 \cdots m_n$ alternative ways.

EXAMPLE 1.38 Three-place numbers are to be formed by applying the digits 1 to 5, inclusive. If there is no restriction on the number of times a given digit can appear, how many such numbers can be formed?

SOLUTION Since each position in the number can be filled in 5 alternative ways, the quantity of such numbers is $5^3 = 125$.

EXAMPLE 1.39 Four-place numbers are to be formed according to the following specifications: Proceeding from left to right, the first digit can range from 1 to 5, inclusive; the second digit can range from 0 to 8, inclusive; the third digit can be 3 or 7; the fourth digit must be 0. How many numbers can be formed?

SOLUTION The quantity of numbers is

$$5 \times 9 \times 2 \times 1 = 90$$

In the subsequent material, we shall apply factorial numbers, which are defined in Art. 1.14.

An arrangement of a set of items or group of individuals in which the order or rank is significant is called a *permutation.* The arrangement may contain the entire set of items or only part of the set. For example, the following are permutations of the first 3 letters of the alphabet taken 2 at a time: *ab, ba, ac, ca, bc, cb.*

To form a permutation of items, we must answer two questions: Which items shall we select? In what order shall we arrange them? To form a permutation of individuals, we must answer these questions: Which individuals shall we select? In what manner shall we assign ranks (captain, cocaptain, etc.) to them?

EXAMPLE 1.40 How many permutations can be formed of the first 7 letters of the alphabet taken 3 at a time?

SOLUTION The first position can be assigned to any one of 7 letters. When the assignment has been made, the number of available letters has been reduced to 6.

Therefore, the second position can be assigned to any one of 6 letters, and the third position can then be assigned to any one of 5 letters. Therefore, by Theorem 1.6, the number of possible permutations is $7 \cdot 6 \cdot 5 = 210$.

Let $P(n,r)$ denote the number of possible permutations of n items taken r at a time. By generalizing from Example 1.40, we obtain

$$P(n,r) = n(n-1)(n-2)\cdots(n-r+1) = \frac{n!}{(n-r)!} \tag{1.21}$$

In the special case where $r = n$, we have

$$P(n,n) = n! \tag{1.21a}$$

EXAMPLE 1.41 A committee consisting of a president, treasurer, and secretary is to be formed. Eight individuals are available for selection, and they are all qualified for each of the three offices. In how many ways can the committee be formed?

SOLUTION Since the committee numbers will have distinctive ranks, each committee represents a *permutation* of 8 individuals taken 3 at a time. Therefore, the number of possible committees is $8!/5! = 8 \cdot 7 \cdot 6 = 336$.

Assume that we have formed all permutations of the 8 characters in the word *parabola*, taken all at a time, and let X denote the number of permutations in the set. If we make the 3 *a*'s distinguishable in some manner and then rearrange them, we transform each permutation in the original set to 3! permutations in a new set. Since the new set contains 8! permutations, it follows that $3!X = 8!$, and $X = 8!/3! = 6720$.

Let $P(n,n,k \text{ alike})$ denote the number of permutations of n items taken n at a time, where k items in the set are indistinguishable from one another. From the foregoing illustrative case, we obtain

$$P(n,n,k \text{ alike}) = \frac{n!}{k!} \tag{1.22}$$

This equation can be extended to cases where the set of n items contains several subsets of indistinguishable items.

EXAMPLE 1.42 How many permutations can be formed by using the characters in the word *dissatisfied*, taken all at a time?

SOLUTION The word contains 12 characters; of these, 2 are *d*, 3 are *i*, and 3 are *s*. By an extension of Eq. (1.22), we have

$$\text{No. of permutations} = \frac{12!}{2!\,3!\,3!} = 6{,}652{,}800$$

A *circular* (or *ring*) *permutation* is one in which items are arranged in a closed loop and we are concerned solely with the *relative order* of the items. Thus, the permutation has no specific beginning. Since a circular permutation consisting of r items can be cut at any one of r points and then straightened to form a *linear* permutation, it follows that there are r possible linear permutations corresponding to every circular permutation.

Let $CP(n,r)$ denote the number of possible circular permutations of n items taken r at a time. From the foregoing discussion, we obtain

$$CP(n,r) = \frac{n!}{r[(n-r)!]} \tag{1.23}$$

EXAMPLE 1.43 A group consists of 6 individuals. Four individuals will be selected at random and seated at a round table. If it is only the relative order of these individuals that is significant, how many seating arrangements can be devised?

SOLUTION

$$\text{No. of arrangements} = \frac{6!}{4(2!)} = 90$$

1.24 COMBINATIONS

A grouping of items or individuals in which the order or rank is of no significance, or in which the order or rank is predetermined, is termed a *combination.* To form a permutation, we must answer two questions; these are presented in Art. 1.23. By contrast, to form a combination, we must answer only one question: Which items or individuals shall we select?

Let $C(n,r)$ denote the number of possible combinations of n items taken r at a time. The r items in a combination can be arranged to form $r!$ permutations, and it follows that $(r!)C(n,r) = P(n,r)$. Therefore,

$$C(n,r) = \frac{n!}{r!(n-r)!} = \binom{n}{r} \tag{1.24}$$

(The expression at the right is a binomial coefficient. Refer to Art. 1.16.) For mathematical consistency, we set $C(n,0) = 1$.

EXAMPLE 1.44 Five-place numbers are to be formed by using the digits 1 to 8, inclusive, with the digits arranged in descending order of magnitude. For example, 76431 is acceptable but 76413 is unacceptable. How many such numbers can be formed?

SOLUTION Since the order of the digits is predetermined, each number represents a *combination* of 8 digits taken 5 at a time. The quantity of such numbers is

$$\frac{8!}{5!3!} = 56$$

EXAMPLE 1.45 A slate of 8 officers is to be formed. One officer will be president, a second vice president, and a third treasurer. If 10 individuals are available for selection and they are all equally qualified, in how many ways can the slate be formed?

SOLUTION The three titled officers will form a *permutation,* and the five untitled officers will form a *combination.* Thus, each prospective slate is a composite of a permutation and a combination.

We shall first form the permutation. The number of possibilities is $P(10,3) = 10 \cdot 9 \cdot 8 = 720$. Seven individuals are now available to fill the five untitled positions, and the number of possibilities is

$$C(7,5) = \frac{7!}{5!\,2!} = \frac{7 \cdot 6}{2} = 21$$

By the law of multiplication, the number of ways in which the slate can be formed is $720 \times 21 = 15{,}120$.

Alternatively, we can reverse the sequence by first forming the combination and then forming the permutation. The calculations become $C(10,5) = 252$ and $P(5,3) = 60$, and the number of possible slates is $252 \times 60 = 15{,}120$.

As a third possibility, we can first select the eight committee members and then select the three titled members from this group. The calculations become $C(10,8) = 45$ and $P(8,3) = 336$, and the number of possible slates is $45 \times 336 = 15{,}120$.

We shall now consider a situation in which a set of objects is to be *partitioned.* Assume the following: We have a set of n distinguishable objects; this set is to be divided into r subsets, where $r < n$; the size of each subset is predetermined. If our interest centers solely about the identity of the objects in a subset and not their arrangement, these objects constitute a *combination.*

EXAMPLE 1.46 A set consists of 12 objects, and these objects are to be placed in three cells: A, B, and C. The number of objects that a cell will contain is as follows: cell A, 7; cell B, 2; cell C, 3. How many different placements are possible?

SOLUTION Let N denote this number. The number of ways in which cell A can be filled is $C(12,7)$. After objects have been placed in cell A, there remain $12 - 7 = 5$ objects. Therefore, the number of ways in which cell B can be filled is $C(5,2)$. Similarly, the number of ways in which cell C can be filled is $C(3,3)$. By the law of multiplication,

$$N = C(12,7) \times C(5,2) \times C(3,3)$$

$$= \frac{12!}{7!\,5!}\,\frac{5!}{2!\,3!}\,\frac{3!}{3!} = \frac{12!}{7!\,2!\,3!} = 7920$$

EXAMPLE 1.47 A firm has 14 machines to be serviced, and three crews of employees are available. If 9 machines are to be assigned to crew A, 3 machines to crew B, and 2 machines to crew C, how many work assignments are possible?

SOLUTION Proceeding as in Example 1.46, we obtain

$$N = \frac{14!}{9!\,3!\,2!} = 20{,}020$$

Refer to Eq. (1.17*a*) in Art. 1.17. By combining this equation with Eq. (1.24) and then setting $a = b = 1$, we obtain the following:

$$C(n,0) + C(n,1) + C(n,3) + \cdots + C(n,n) = 2^n \tag{1.25}$$

There are many situations where we have a group of objects and we must form every possible type of combination of these objects. Equation (1.25) reveals how many such combinations exist. Example 2.3 provides an illustration.

1.25 PERMUTATIONS AND COMBINATIONS SUBJECT TO RESTRICTIONS

Where restrictions are imposed on the manner in which a permutation or combination may be formed, the number of acceptable permutations or combinations can be found by simple logic.

EXAMPLE 1.48 A committee consisting of 5 members is to be formed, and 12 individuals are available for assignment to the committee. The committee members will have equal rank. In how many ways can the committee be formed if Smith and Jones cannot both serve on the committee?

SOLUTION Since the committee members will have equal rank, each committee represents a *combination.* In the absence of any restrictions, the number of possible committees would be

$$C(12,5) = \frac{12!}{5!\,7!} = \frac{12 \cdot 11 \cdot 10 \cdot 9 \cdot 8}{5 \cdot 4 \cdot 3 \cdot 2} = 792$$

We shall now determine the number of committees that violate the imposed restriction. Consider that Smith and Jones are both appointed. Three committee members remain to be appointed, and 10 individuals are available. Therefore, the number of unacceptable committees is

$$C(10,3) = \frac{10!}{3!\,7!} = \frac{10 \cdot 9 \cdot 8}{3 \cdot 2} = 120$$

Then

$$\text{No. acceptable committees} = 792 - 120 = 672$$

EXAMPLE 1.49 Six individuals are to be seated at a round table, but Adams and Barnes must be seated alongside each other. If we are concerned solely with the relative order of these individuals, how many seating arrangements can be devised?

SOLUTION Consider that we first assign a seat to Adams. We can then place Barnes either at his right or at his left. Four seats remain to be assigned, and the number of possible arrangements is $P(4,4) = 4! = 24$. By Theorem 1.6 (the law of multiplication), the number of possible seating arrangements is $2 \times 24 = 48$.

EXAMPLE 1.50 Five-place numbers are to be formed by applying the digits 1 to 8, inclusive. The following restrictions apply: Each digit can appear only once in the number; if the digits 4 and 7 both appear, then 7 must occupy the second position to the right of 4. For example, the numbers 84572 and 61437 are acceptable. How many such numbers can be formed?

SOLUTION The group of acceptable numbers can be divided into four sets. Set 1 consists of the numbers that contain neither 4 nor 7; set 2 consists of those that contain 4 but not 7; set 3 consists of those that contain 7 but not 4; set 4 consists of those that contain both 4 and 7. We shall calculate the number of permutations associated with each set.

Set 1. In forming the number, we have six digits available and five positions to be filled. The number of permutations is

$$P(6,5) = 720$$

Set 2. We place 4 in any one of five positions. We now have six digits available and four positions to be filled. The number of permutations is

$$5 \times P(6,4) = 5 \times 360 = 1800$$

Set 3. The number of permutations is 1800.

Set 4. We place 4 in either of the first three positions and we then place 7 in the second position to the right of 4. We now have six digits available and three positions to be filled. The number of permutations is

$$3 \times P(6,3) = 3 \times 120 = 360$$

Therefore, the quantity of numbers that can be formed is

$$720 + 2 \times 1800 + 360 = 4680$$

An alternative method of solution consists of these steps: First, calculate the quantity of numbers if the second restriction were absent (i.e., if 4 and 7 could have any relative order). This quantity is $P(8,5) = 6720$. Second, determine the quantity of numbers that violate this restriction. This quantity is found to be 2040. Therefore, the quantity of acceptable numbers is $6720 - 2040 = 4680$.

1.26 WORK PROBLEMS

We shall now investigate a type of problem in which a group of individuals is working on a project and it is necessary to compute some specified quantity. In this type of problem, it is understood that the rate at which a particular individual performs work remains constant as time elapses and that this rate is independent of the size of the group.

EXAMPLE 1.51 Fourteen mechanics can produce 21 units of a commodity in 5 h. How many mechanics are required to produce 54 units of the commodity in 9 h?

SOLUTION It is understood that all mechanics perform work at the same rate. A problem of this type can be solved by multiple methods. We shall solve it by a method that makes use of proportions, as this is the least time-consuming one. Let n denote the number of mechanics required. This quantity is directly proportional to the number of units to be produced and inversely proportional to the allotted time. Then

$$n = 14\left(\frac{54}{21}\right)\left(\frac{5}{9}\right) = 20 \text{ mechanics}$$

EXAMPLE 1.52 Three individuals, A, B, and C, are each capable of completing a job. If they work singly, the time required for completion is as follows: A, 20 h; B, 16 h; C, 15 h. If the three individuals work together, how long will it take them to complete the job?

SOLUTION Let t denote the time required to complete the job in hours. When these individuals work singly, the fractional part of the job that is completed in 1 h is 1/20 for A, 1/16 for B, and 1/15 for C. Therefore, when they work together for time t, we have

$$\frac{t}{20} + \frac{t}{16} + \frac{t}{15} = 1$$

$$t = 5.58 \text{ h}$$

EXAMPLE 1.53 With reference to Example 1.52, A and B will work together for 3 h, and C will then complete the job. How long must C work?

SOLUTION Let t denote the time that C works. Then

$$\frac{3}{20} + \frac{3}{16} + \frac{t}{15} = 1$$

$$t = 9.94 \text{ h}$$

EXAMPLE 1.54 Smith can complete a job in 7 h, and Jones can complete it in 6 h. They work together for 2 h, and Smith leaves. How long will it take Jones to complete the job?

SOLUTION Let t denote this time in hours. When they work singly, the fractional part of the job that is completed in 1 h is 1/7 for Smith and 1/6 for Jones. The total number of hours worked is 2 for Smith and $2 + t$ for Jones. Then

$$2\left(\frac{1}{7}\right) + (2 + t)\left(\frac{1}{6}\right) = 1$$

Solving, we obtain $t = 16/7 = 2.29$ h.

EXAMPLE 1.55 A firm digs trenches of uniform width and then backfills. A crew of 6 laborers can dig a 45-m trench in 9 h, and a crew of 5 laborers can backfill a 35-m trench in 3 h. How long will it take a crew of 8 laborers to dig and backfill a 55-m trench?

SOLUTION Let t_d and t_b denote the time required for digging and for backfill, respectively. Each of these quantities is directly proportional to the depth of the trench and inversely proportional to the size of the crew. Then

$$t_d = 9\left(\frac{55}{45}\right)\left(\frac{6}{8}\right) = 8.25 \text{ h}$$

$$t_b = 3\left(\frac{55}{35}\right)\left(\frac{5}{8}\right) = 2.95 \text{ h}$$

The total time required is $8.25 + 2.95 = 11.20$ h.

CHAPTER 2
SET THEORY

2.1 DEFINITIONS

A collection of related items is referred to as a *set*, and the items that compose the set are referred to as its *elements* or *members*. Sets are highly useful in the development of logical thinking, and set theory is closely related to boolean algebra.

For identification, a set is assigned an uppercase letter and an element is assigned a lowercase letter. If q is an element of set A, the information is recorded symbolically in this manner:

$$q \in A$$

On the other hand, if q is not an element of A, the notation is $q \notin A$. We shall assume that no two items in a set are identical.

A set may be finite or infinite. Thus, a set that consists of the letters of the alphabet is finite, and a set that consists of all positive integers is infinite.

If a set is relatively small, it can be described very simply by exhibiting the elements. They are recorded in a row, and the row is enclosed with braces. For example, assume that set A consists of the first five letters of the alphabet. The description assumes this form:

$$A = \{a, b, c, d, e\}$$

The sequence in which the elements are recorded is immaterial. If a set is relatively large, it is described by expressing the rule by which elements were selected for inclusion. For example, the notation

$$B = \{x : x \text{ is a prime number, } x > 11\}$$

states that B is a set consisting of all prime numbers greater than 11.

If sets A and B are identical, they are said to be *equal* to each other, and the symbolic expression is $A = B$. On the other hand, if A and B are not identical, the notation is $A \neq B$.

It is convenient to define a *null* or *empty set*. This is a set that contains no elements, and it is denoted by $\varnothing$. For example, if the oldest individual living in a certain town is 95 years of age, a set consisting of individuals in this town who are 100 years of age or older is a null set.

Associated with a given set is a *universal set*, denoted by U. It consists of both those items that are in the given set and those items that are related to them in some defined manner. Thus, if set A consists of the families in a certain town who own dogs, we define U as the set that consists of the families in this town who own *pets*.

2.2 VENN DIAGRAMS AND RELATIONSHIPS BETWEEN SETS

In the analysis of sets, it is highly advantageous to represent each set in graphical form. This is accomplished by drawing a closed geometric shape, such as a circle or rectangle, and considering that each element in the set is represented by a point that lies within the enclosed space. Thus, this space represents the set, and it is labeled accordingly. A diagram of this type is termed a *Venn diagram.* It is unnecessary to draw the diagram to any scale, as it is intended simply as an aid to visualization.

Figure 2.1 exhibits three possible relationships between sets A and B. In *a*, set A lies entirely within set B. Therefore, every element of A is also an element of B. In *b*, sets A and B overlap. Therefore, some elements belong to both A and B. In *c*, sets A and B are completely distinct. Therefore, they have no elements in common.

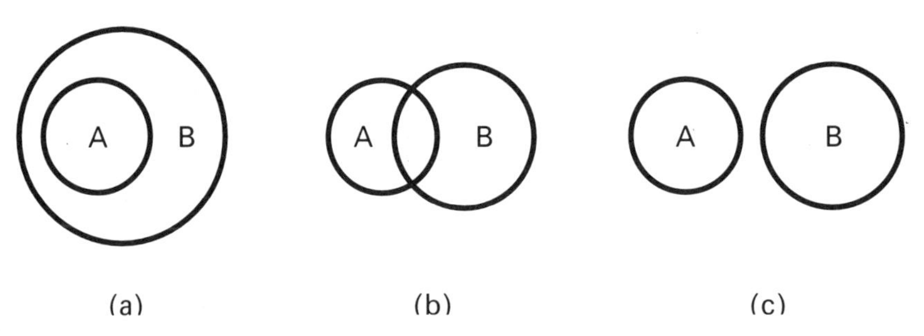

FIGURE 2.1 Venn diagrams. (*a*) Subset and superset; (*b*) overlapping sets; (*c*) disjoint sets.

With reference to Fig. 2.1*a*, set A is called a *subset* of set B, and B is called a *superset* of A. The symbolic expression is as follows:

$$A \subset B \qquad \text{and} \qquad B \supset A$$

For example, if A consists of all even integers from 10 to 80 and B consists of all even integers from 0 to 100, then $A \subset B$.

By convention, a set is considered to be a subset of itself. Therefore, if $A \subset B$, two possibilities exist: A and B are identical, or B contains elements that are not in A. In the latter case, A is said to be a *proper subset* of B. As an illustration, consider that A is the set of chemical engineers employed by a firm and B is the set of chemical, mechanical, and petroleum engineers employed by this firm. Then A is a proper subset of B. From the definition, it follows that the null set is a proper subset of all other sets.

If A is not a subset of B, the notation is $A \not\subset B$.

Theorem 2.1. If A is a subset of B and B is a subset of C, then A is a subset of C.

The sets A and B in Fig. 2.1*b* are called *overlapping* or *intersecting sets*, and the sets in Fig. 2.1*c* are said to be *disjoint* with respect to each other.

EXAMPLE 2.1 Sets A, B, C, and D are composed of the following: A, all prime numbers from 3 to 79, inclusive; B, all integers from 1 to 100, inclusive; C, all

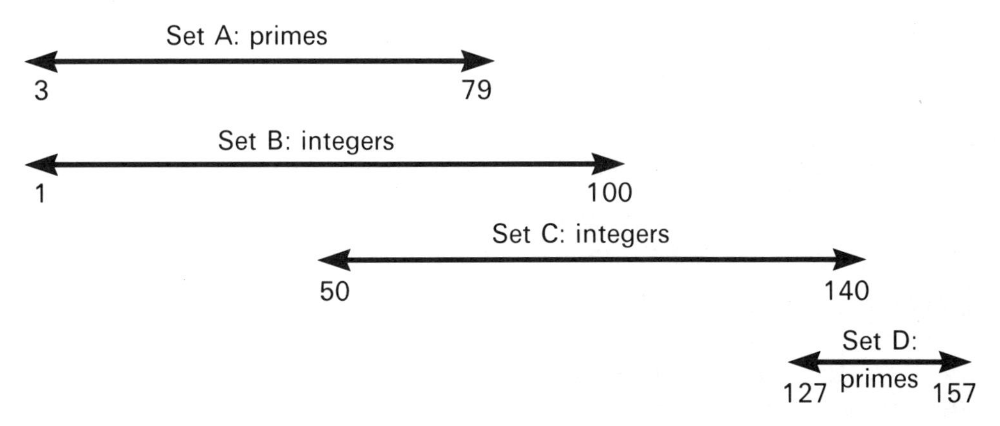

FIGURE 2.2 Ranges, contents, and relative positions of sets.

integers from 50 to 140, inclusive; D, all prime numbers from 127 to 157, inclusive. How are these sets related to one another? Draw the Venn diagram.

SOLUTION Refer to Fig. 2.2, which displays the extent, composition, and relative position of each set. Set A is a subset of B. Sets A and C overlap because they have elements in common (namely, 53, 59, 61, 67, 71, 73, and 79). Sets B and C overlap because they have elements in common (namely, the integers from 50 to 100, inclusive). Sets C and D overlap because they have elements in common (namely, 127, 131, 137, and 139). Set D is disjoint with respect to both A and B. The Venn diagram appears in Fig. 2.3.

2.3 NUMBER OF SUBSETS

Consider that a set consists of n items and that we are to form a subset that consists of r items, where $r \leq n$. The items in the subset constitute a *combination* of the n items in the set taken r at a time. Therefore, the number of subsets that can be formed is $C(n,r)$. (Refer to Art. 1.24.)

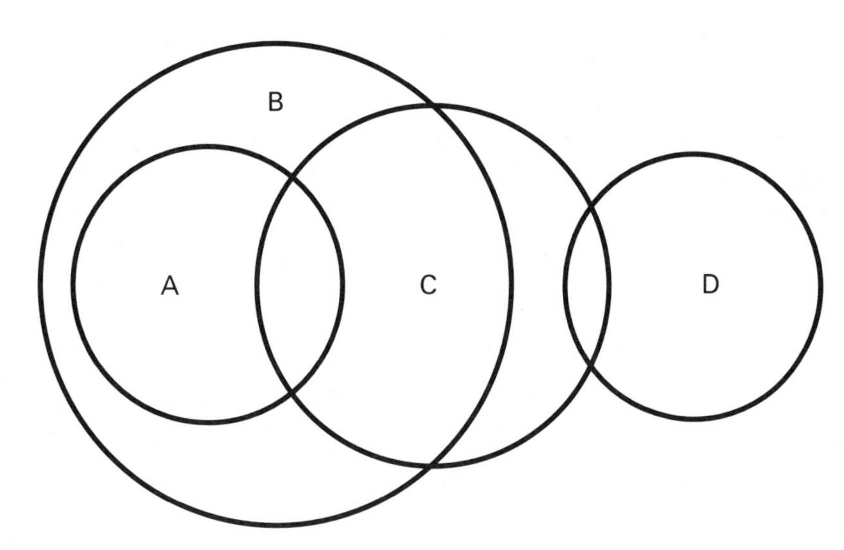

FIGURE 2.3 Venn diagram.

EXAMPLE 2.2 Set A contains 12 items, and we are required to form subsets that contain 9 items. How many such subsets can be formed?

SOLUTION The number of possible subsets is

$$C(12,9) = \frac{12!}{9!\,3!} = 220$$

Equation (1.25) yields the following principle:

Theorem 2.2. Let A denote a set that consists of n items. If the null set and set A itself are included, the number of subsets that are inherent in A is 2^n.

EXAMPLE 2.3 Set A consists of the first three letters of the alphabet. How many subsets of A can be formed? Record the subsets.

SOLUTION The number of possible subsets is $2^3 = 8$. The subsets are as follows:

$$\begin{array}{cccc} \varnothing & \{a\} & \{b\} & \{c\} \\ \{a, b\} & \{a, c\} & \{b, c\} & \{a, b, c\} \end{array}$$

The subsets that are inherent in a given set A constitute the *power set* of A. Thus, the solution to Example 2.3 exhibits the power set corresponding to the given set.

2.4 OPERATIONS WITH SETS

When we are given two sets, we can operate on them in some prescribed manner to form a third set, which we shall term a *constructed set*. The given sets will be called A and B and the constructed set will be called C. Sets A and B are represented in Fig. 2.4, where we have assumed that the sets overlap. Set C is shown shaded. The operations are as follows:

1. Union. Consider that C is formed by *combining* sets A and B. Then C is called the *union* of A and B, and the symbolic expression is

$$C = A \cup B$$

(Some books use the notation $C = A + B$.) Thus, an item is an element of C if it is an element of either A or B (or both). Refer to Fig. 2.4*a*.

The symbolic description of C is

$$C = \{x : x \in A \text{ or } x \in B\}$$

EXAMPLE 2.4 The following sets are given:

$$A = \{1, 3, 5, 7, 9\} \qquad \text{and} \qquad B = \{1, 2, 3, 4, 5\}$$

If $C = A \cup B$, construct C.

SOLUTION It is helpful to exhibit the sets in the tabular form shown below, where an X signifies that the number at the top of the column is an element of the set.

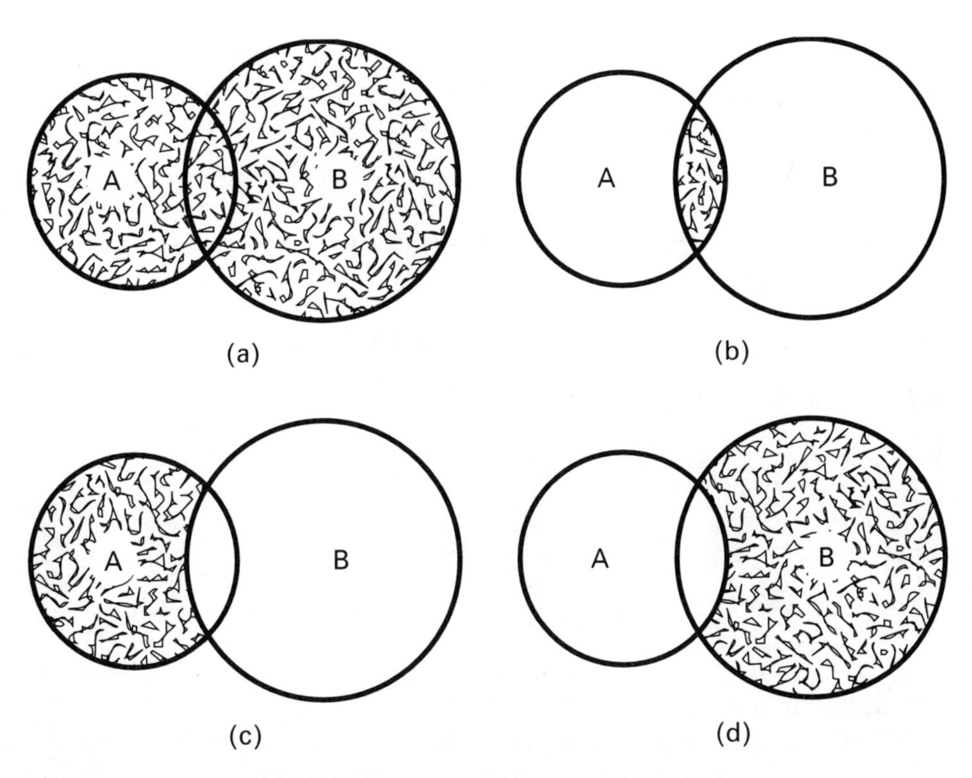

FIGURE 2.4 Representation of constructed set by shading. (*a*) $A \cup B$; (*b*) $A \cap B$; (*c*) $A - B$; (*d*) $B - A$.

	1	2	3	4	5	6	7	8	9
A	X		X		X		X		X
B	X	X	X	X	X				

$$C = \{1, 2, 3, 4, 5, 7, 9\}$$

2. ***Intersection.*** Consider that C is formed by selecting only those elements that lie in *both* A and B. Then C is called the *intersection* of A and B, and the symbolic expression is

$$C = A \cap B$$

(Some books use the notation $C = AB$ or $C = A \times B$.) Refer to Fig. 2.4*b*. The symbolic description of C is

$$C = \{x : x \in A \text{ and } x \in B\}$$

EXAMPLE 2.5 With reference to sets A and B in Example 2.4, construct C if $C = A \cap B$.

SOLUTION

$$C = \{1, 3, 5\}$$

*3. **Difference.*** Consider that C is formed by taking those elements that lie in A but excluding those that lie in B. Then C is called the *difference* between A and B, and the symbolic expression is

$$C = A - B$$

(Some books use the notation $A \backslash B$.) Refer to Fig. 2.4*c*. The symbolic description of C is

$$C = \{x : x \in A \text{ and } x \notin B\}$$

Figure 2.4*d* represents the set $C = B - A$.

EXAMPLE 2.6 With reference to sets A and B in Example 2.4, construct sets C and D if $C = A - B$ and $D = B - A$.

SOLUTION

$$C = \{7,9\} \qquad D = \{2,4\}$$

The difference between the universal set U and a given set A is called the *complement* of A, and it is denoted by A^c. (Some books denote the complement by A' or $\bar{A}$.) For example, if A consists of all positive even integers and U consists of all positive integers, then A^c consists of all positive odd integers. The notation $(A^c)^c$ means the complement of A^c. Manifestly, $(A^c)^c = A$.

EXAMPLE 2.7 Identify the condition under which each of the following statements is valid: (*a*) $A \cup B = A$; (*b*) $A \cap B = \varnothing$; (*c*) $A \cap B = A$; (*d*) $A - B = A$; (*e*) $B - A = \varnothing$.

SOLUTION As an aid in interpreting each equation, we shall refer to the appropriate Venn diagram in Fig. 2.4 and consider that one circle is displaced until the stipulated condition arises.

Part a. Since the addition of B to A adds nothing to A, it follows that B is a subset of A. In Fig. 2.4*a*, consider that circle B is not larger than circle A and that B is displaced to the left until it lies completely within A.

Part b. The given equation states in effect that there are no elements that are common to A and B. Therefore, A and B are disjoint sets. In Fig. 2.4*b*, consider that circle B is displaced to the right until it lies completely beyond circle A.

Part c. Since the elements that are common to A and B all lie in A, it follows that A is a subset of B. In Fig. 2.4*b*, consider that circle A is not larger than circle B and that A is displaced to the right until it lies completely within B.

Part d. Since the subtraction of B from A leaves A unchanged, it follows that A and B are disjoint. In Fig. 2.4*c*, consider that circle B is displaced to the right until it lies completely beyond A.

Part e. Since the subtraction of A from B leaves a null set, it follows that B is a subset of A. In Fig. 2.4*d*, consider that circle B is not larger than circle A and that B is displaced to the left until it lies completely within A.

2.5 LAWS OF SET OPERATIONS

There are various laws governing the operations with sets. We shall not record these laws in detail, but we shall illustrate how they can be proved readily by the use of

Venn diagrams. Example 2.8 has a two-fold purpose: to prove a particular law, and to provide practice in recognizing a constructed set. In the Venn diagram, we shall use shading to identify the area that represents the constructed set.

EXAMPLE 2.8 Prove the following:

$$(A \cup B) \cap (A \cup C) = A \cup (B \cap C)$$

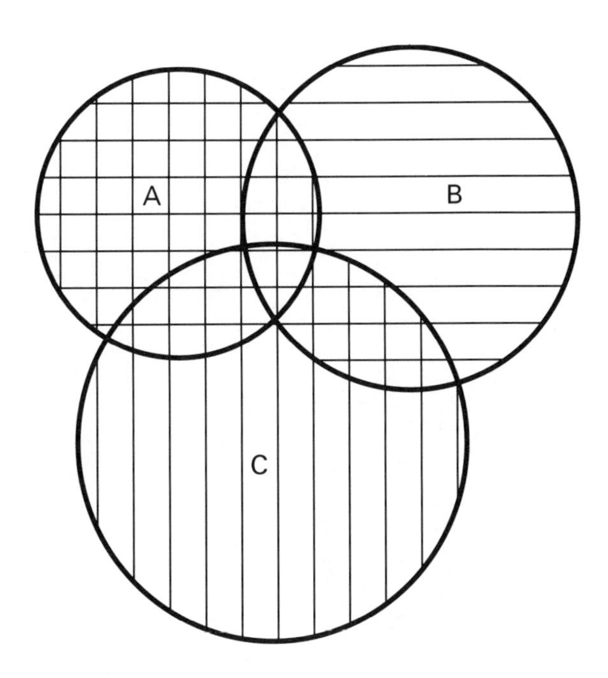

FIGURE 2.5

SOLUTION Let D denote the set having the expression to the left of the equals sign, and refer to Fig. 2.5. Any point that lies within circle A or B satisfies $A \cup B$. Therefore, draw *horizontal* lines through both A and B. Similarly, any point that lies within circle A or C satisfies $A \cup C$. Therefore, draw *vertical* lines through A and C. Set D is represented by the area that has *both* horizontal and vertical lines. Thus, D encompasses the area within A and the area that is common to B and C. The given equation is thus proved.

CHAPTER 3
MATRIX ALGEBRA

3.1 DEFINITIONS AND NOTATION

A *matrix* is a rectangular array of numbers (or of letters that represent numbers). The array is enclosed in brackets or parentheses, and each matrix is assigned an uppercase boldface letter for identification.

The numbers that compose the matrix are termed its *elements* or *members*. Elements that lie on a horizontal line constitute a *row*, and those that lie on a vertical line constitute a *column*. The rows and columns are assigned identifying numbers by the following convention: Number the rows from top to bottom; number the columns from left to right. A row or column is described as *nonzero* if it contains at least one nonzero element.

The *size* of a matrix is referred to as its *order* or *dimension*. A matrix having m rows and n columns is said to be of order $m \times n$, or it is described as an $m \times n$ matrix.

EXAMPLE 3.1 The following matrix is given:

$$\mathbf{A} = \begin{bmatrix} 7 & 5 & 6 & 3 \\ -11 & 8 & -3 & 2 \\ 4 & 9 & 0 & -15 \end{bmatrix}$$

Specify the size of the matrix, and identify the third row and second column.

SOLUTION Since the matrix contains 3 rows and 4 columns, it is a 3×4 matrix. The third row is 4, 9, 0, -15; the second column is 5, 8, 9.

Where the elements of a matrix are referred to by algebraic symbols, the practice is as follows: Each element is assigned the lowercase italicized letter corresponding to the matrix label and two subscripts are appended. The first subscript is the row number; the second is the column number. For example, the symbol b_{25} (read "*b* sub two five") identifies the element in matrix **B** that lies in the second row and fifth column. Thus, with reference to the matrix in Example 3.1, we have $a_{13} = 6$ and $a_{32} = 9$.

A matrix that consists of a single row is called a *row vector*, and one that consists of a single column is called a *column vector*. A matrix in which all elements are zero is termed a *null matrix*, and it is designated as **0**.

Matrices **A** and **B** are *equal* to each other if they are of the same order and all corresponding elements are equal. The equality is expressed by the notation $\mathbf{A} = \mathbf{B}$.

Consider that one or more rows or columns of a matrix **A** are deleted, and let **B** denote the matrix that remains. Then **B** is a *submatrix* of **A**.

EXAMPLE 3.2 The following matrix is given:

$$\mathbf{A} = \begin{bmatrix} -3 & 2 & 9 & -12 \\ 11 & 7 & -4 & 1 \\ 5 & 8 & 6 & 3 \end{bmatrix}$$

A submatrix **B** is formed by deleting the third row and the second and fourth columns. Identify **B**.

SOLUTION

$$\mathbf{B} = \begin{bmatrix} -3 & 9 \\ 11 & -4 \end{bmatrix}$$

Consider that a given matrix **A** is transformed by converting its rows to columns and its columns to rows. Thus, a_{ij} becomes a_{ji}. The matrix that results is known as the *transpose* of **A**, and it is denoted by $\mathbf{A}^T$. For example, if

$$\mathbf{A} = \begin{bmatrix} 3 & 9 \\ 2 & 6 \\ 4 & 1 \\ 8 & 5 \end{bmatrix} \quad \text{then} \quad \mathbf{A}^T = \begin{bmatrix} 3 & 2 & 4 & 8 \\ 9 & 6 & 1 & 5 \end{bmatrix}$$

Now consider that a given matrix **A** is transformed by changing the algebraic sign of each element. The matrix that results is the *negative* of **A**, and it is denoted by $-\mathbf{A}$.

3.2 SQUARE MATRICES

A matrix in which the number of rows equals the number of columns is termed a *square matrix*. If n denotes this number, the matrix is said to be of the nth order, or of order n.

The diagonal line that extends from the upper-left corner to the lower-right corner of a square matrix is called the *principal* or *main diagonal.* Let **A** denote a square matrix. The elements $a_{11}, a_{22}, \ldots, a_{nn}$ that lie on this diagonal are termed the *diagonal elements.* The sum of the diagonal elements is the *trace* of the matrix.

Names have been assigned to special types of square matrices, and the following will serve as illustrations:

$$\mathbf{A} = \begin{bmatrix} 9 & 5 & 4 & 2 \\ 0 & 1 & 8 & 0 \\ 0 & 0 & -2 & 3 \\ 0 & 0 & 0 & 7 \end{bmatrix} \qquad \mathbf{B} = \begin{bmatrix} 9 & 0 & 0 & 0 \\ 0 & 4 & 0 & 0 \\ 0 & 0 & 8 & 0 \\ 0 & 0 & 0 & 7 \end{bmatrix}$$

Triangular Diagonal

$$\mathbf{C} = \begin{bmatrix} 6 & 0 & 0 \\ 0 & 6 & 0 \\ 0 & 0 & 6 \end{bmatrix} \qquad \mathbf{D} = \mathbf{I} = \begin{bmatrix} 1 & 0 & 0 \\ 0 & 1 & 0 \\ 0 & 0 & 1 \end{bmatrix}$$

Scalar Unit (identity)

$$\mathbf{E} = \begin{bmatrix} 2 & 6 & 3 \\ 6 & 9 & 8 \\ 3 & 8 & 4 \end{bmatrix} \qquad \mathbf{F} = \begin{bmatrix} 0 & -6 & 3 \\ 6 & 0 & -8 \\ -3 & 8 & 0 \end{bmatrix}$$

Symmetric Skew-symmetric

A matrix such as **A** in which all elements above or below the principal diagonal are zero is a *triangular matrix*. Moreover, the matrix is *upper-triangular* or *lower-triangular*, respectively, according to whether the nonzero elements lie above or below the principal diagonal. Thus, **A** is an upper-triangular matrix.

A matrix such as **B** in which all nondiagonal elements are zero is a *diagonal matrix*. Thus, $b_{ij} = 0$ if $i \neq j$. A diagonal matrix such as **C** in which all diagonal elements are equal is a *scalar matrix*.

A scalar matrix such as **D** in which all diagonal elements are 1 is a *unit* or *identity matrix*, and it is denoted by **I**. A matrix such as **E** in which $e_{ij} = e_{ji}$ is termed a *symmetric matrix*. Corresponding rows and columns are identical, and therefore $\mathbf{E}^T = \mathbf{E}$.

A matrix such as **F** in which $f_{ij} = -f_{ji}$ is termed a *skew-symmetric or anti-symmetric matrix*. The diagonal elements must perforce be zero, and $-\mathbf{F}^T = \mathbf{F}$.

3.3 ADDITION AND SUBTRACTION OF MATRICES

Let **A** and **B** denote two matrices of identical order. Consider that corresponding elements are added to form a third matrix **C**. Then **C** is the *sum* of **A** and **B**, and we write $\mathbf{C} = \mathbf{A} + \mathbf{B}$. Thus, the defining equation of addition is

$$c_{ij} = a_{ij} + b_{ij}$$

where i and j assume all possible values.

Analogously, matrix *subtraction* entails the subtraction of corresponding elements. Let $\mathbf{D} = \mathbf{A} - \mathbf{B}$. Then

$$d_{ij} = a_{ij} - b_{ij}$$

and **D** is called the *difference* between **A** and **B**.

EXAMPLE 3.3 The following matrices are give:

$$\mathbf{A} = \begin{bmatrix} 3 & 6 \\ 5 & 8 \\ -2 & 9 \end{bmatrix} \qquad \mathbf{B} = \begin{bmatrix} -6 & 1 \\ 0 & 9 \\ 8 & 3 \end{bmatrix}$$

Construct matrices **C** and **D** such that $\mathbf{C} = \mathbf{A} + \mathbf{B}$ and $\mathbf{D} = \mathbf{A} - \mathbf{B}$.

SOLUTION

$$\mathbf{C} = \begin{bmatrix} -3 & 7 \\ 5 & 17 \\ 6 & 12 \end{bmatrix} \qquad \mathbf{D} = \begin{bmatrix} 9 & 5 \\ 5 & -1 \\ -10 & 6 \end{bmatrix}$$

3.4 MULTIPLICATION AND DIVISION OF A MATRIX BY A SCALAR

A real number is often referred to as a *scalar quantity*, or simply a *scalar*, to distinguish it clearly from a matrix. Consider that all elements of matrix **A** are multiplied by the scalar α to form a matrix **B**. This operation is called the *multiplication* of **A** and α, and it is expressed symbolically as $\mathbf{B} = \alpha\mathbf{A}$. Thus, the defining equation for **B** is

$$b_{ij} = \alpha a_{ij}$$

The *division* of a matrix by a scalar is defined in an analogous manner.

EXAMPLE 3.4 The following matrix is given:

$$\mathbf{A} = \begin{bmatrix} 8 & -2 & 14 \\ 6 & 16 & 12 \end{bmatrix}$$

Construct matrices **B** and **C** such that $\mathbf{B} = 3\mathbf{A}$ and $\mathbf{C} = \mathbf{A}/2$.

SOLUTION

$$\mathbf{B} = \begin{bmatrix} 24 & -6 & 42 \\ 18 & 48 & 36 \end{bmatrix} \qquad \mathbf{C} = \begin{bmatrix} 4 & -1 & 7 \\ 3 & 8 & 6 \end{bmatrix}$$

3.5 MULTIPLICATION OF MATRICES

Let **A** and **B** denote two matrices that have this characteristic: The number of *columns* in **A** equals the number of *rows* in **B**. These matrices are *conformable* with respect to each other, and they can be combined by an operation called the *multiplication* of matrices. Let **C** denote the resulting matrix. Matrix **C** is the *product* of **A** and **B**, and the operation is expressed symbolically as $\mathbf{C} = \mathbf{AB}$.

The formula for constructing **C** is given by this equation:

$$c_{ij} = a_{i1}b_{1j} + a_{i2}b_{2j} + a_{i3}b_{3j} + \cdots + a_{in}b_{nj} \tag{3.1}$$

where n denotes the number of columns in **A** and the number of rows in **B**, and i and j assume all possible values. Expressed verbally, c_{ij} is formed by taking the ith *row* of **A** and the jth *column* of **B**, multiplying corresponding elements, and summing the products. The formula can be shown symbolically in this manner:

$$c_{ij} = (i\text{th row of } \mathbf{A})(j\text{th column of } \mathbf{B})$$

EXAMPLE 3.5 The following matrices are given:

$$\mathbf{A} = \begin{bmatrix} 4 & 1 & 9 \\ 6 & 2 & 8 \\ 7 & 3 & 5 \\ 11 & 10 & 12 \end{bmatrix} \qquad \mathbf{B} = \begin{bmatrix} 2 & 9 \\ 5 & 12 \\ 8 & 10 \end{bmatrix}$$

Construct matrix **C** such that **C** = **AB**.

SOLUTION Matrix **A** has 3 columns and **B** has 3 rows. Therefore, the matrices are conformable, and $n = 3$. By allowing i and j in Eq. (3.1) to vary from 1 to 4 and from 1 to 2, respectively, we obtain the following results:

$$c_{11} = 4 \times 2 + 1 \times 5 + 9 \times 8 = 85$$
$$c_{12} = 4 \times 9 + 1 \times 12 + 9 \times 10 = 138$$
$$c_{21} = 6 \times 2 + 2 \times 5 + 8 \times 8 = 86$$
$$c_{22} = 6 \times 9 + 2 \times 12 + 8 \times 10 = 158$$
$$c_{31} = 7 \times 2 + 3 \times 5 + 5 \times 8 = 69$$
$$c_{32} = 7 \times 9 + 3 \times 12 + 5 \times 10 = 149$$
$$c_{41} = 11 \times 2 + 10 \times 5 + 12 \times 8 = 168$$
$$c_{42} = 11 \times 9 + 10 \times 12 + 12 \times 10 = 339$$

Then
$$\mathbf{C} = \begin{bmatrix} 85 & 138 \\ 86 & 158 \\ 69 & 149 \\ 168 & 339 \end{bmatrix}$$

Let **A** be an $m \times n$ matrix, **B** be an $n \times s$ matrix, and **C** = **AB**. Then **C** is an $m \times s$ matrix. Now let [**AB**] denote the order of matrix **AB**. We shall display the foregoing relationship symbolically in this manner:

$$[\mathbf{AB}] = [m \times n][n \times s] = [m \times s] \tag{3.2}$$

EXAMPLE 3.6 If **A** is a 10×9 matrix, **C** has twice as many columns as rows, and **ABC** is a square matrix, determine the order of **B** and of **C**. Verify the results.

SOLUTION Let a and b denote, respectively, the number of rows and number of columns in **B**. Equation (3.2) yields

$$[\mathbf{AB}] = [10 \times 9][a \times b] = [10 \times b]$$

Since the two inner numbers must be equal, $a = 9$. Now let c denote the number of rows in **C**. Equation (3.2) now yields

$$[(\mathbf{AB})\mathbf{C}] = [10 \times b][c \times 2c] = [10 \times 2c]$$

Since the two inner numbers must be equal, $b = c$. Since **ABC** is square, $2c = 10$, $c = 5$, and $b = 5$. Thus, **B** is a 9×5 matrix and **C** is a 5×10 matrix.

The proof is as follows:

$$[\mathbf{AB}] = [10 \times 9][9 \times 5] = [10 \times 5]$$

$$[(\mathbf{AB})\mathbf{C}] = [10 \times 5][5 \times 10] = [10 \times 10]$$

EXAMPLE 3.7 If **A** is a 5×8 matrix, **B** has twice as many rows as columns, and **ABC** has three times as many columns as rows, determine the order of **B** and of **C**. Verify the results.

SOLUTION Let a denote the number of columns in **B**. Then

$$[\mathbf{AB}] = [5 \times 8][2a \times a] = [5 \times a]$$

We find that $2a = 8$ and $a = 4$. Now let b and c denote, respectively, the number of rows and number of columns in **C**. Then

$$[(\mathbf{AB})\mathbf{C}] = [5 \times 4][b \times c] = [5 \times c]$$

We find that $b = 4$ and $c = 3 \times 5 = 15$. Thus, **B** is an 8×4 matrix and **C** is a 4×15 matrix.

The proof is as follows:

$$[\mathbf{AB}] = [5 \times 8][8 \times 4] = [5 \times 4]$$

$$[(\mathbf{AB})\mathbf{C}] = [5 \times 4][4 \times 15] = [5 \times 15]$$

EXAMPLE 3.8 Solve for x, y, and z if

$$\begin{bmatrix} 2 & 5 & -3 \\ -8 & 2 & 4 \\ 7 & -9 & -6 \end{bmatrix} \begin{bmatrix} x \\ y \\ z \end{bmatrix} = \begin{bmatrix} -26 \\ 98 \\ -130 \end{bmatrix}$$

SOLUTION Performing the matrix multiplication, we obtain the following system of simultaneous equations:

$$2x + 5y - 3z = -26$$

$$-8x + 2y + 4z = 98$$

$$7x - 9y - 6z = -130$$

The calculator yields the following solution: $x = -7$; $y = 3$; $z = 9$.

Manifestly, it is imperative in matrix multiplication that the matrices to be multiplied be recorded in the proper order, since **AB** and **BA** have entirely disparate meanings. To avoid confusion, it is said that, when we form the product **AB**, matrix **A** is *postmultiplied* by **B** and **B** is *premultiplied* by **A**.

If **A** is a square matrix and **I** is the unit matrix of the same order as **A**, we have

$$\mathbf{AI} = \mathbf{IA} = \mathbf{A} \tag{3.3}$$

The basic laws pertaining to the multiplication of matrices are as follows:

$$\alpha(\mathbf{AB}) = (\alpha\mathbf{A})\mathbf{B} = \mathbf{A}(\alpha\mathbf{B}) \tag{3.4}$$

where α denotes a scalar.

$$\mathbf{A}(\mathbf{BC}) = (\mathbf{AB})\mathbf{C} \tag{3.5}$$

$$(\mathbf{A} + \mathbf{B})\mathbf{C} = \mathbf{AC} + \mathbf{BC} \tag{3.6}$$

3.6 POSITIVE POWERS OF A MATRIX

If a square matrix **A** is multiplied by itself, the product is a square matrix of the same order as **A**, and it is denoted by $\mathbf{A}^2$. If $\mathbf{A}^2$ is postmultiplied by **A**, the product is $\mathbf{A}^3$; etc. However, by Eq. (3.5), the order of multiplication is immaterial in the present instance. Therefore, taking $\mathbf{A}^1 = \mathbf{A}$ as a starting point, we may define $\mathbf{A}^n$ by this recursive formula:

$$\mathbf{A}^n = \mathbf{A}\mathbf{A}^{n-1} = \mathbf{A}^{n-1}\mathbf{A} \tag{3.7}$$

where n is a positive integer. Matrix $\mathbf{A}^n$ is called a *power matrix* of **A**.

From the definition of a power matrix, it follows that

$$\mathbf{A}^m\mathbf{A}^n = \mathbf{A}^n\mathbf{A}^m = \mathbf{A}^{m+n} \tag{3.8}$$

where m and n are positive integers.

EXAMPLE 3.9 The following matrix is given:

$$\mathbf{A} = \begin{bmatrix} 2 & -4 \\ -3 & 5 \end{bmatrix}$$

Construct the matrices $\mathbf{A}^2$, $\mathbf{A}^3$, and $\mathbf{A}^5$.

SOLUTION Let $\mathbf{B} = \mathbf{A}^2$. Then

$$b_{11} = 2 \times 2 + (-4)(-3) = 16 \qquad b_{12} = 2(-4) + (-4)5 = -28$$

$$b_{21} = (-3)2 + 5(-3) = -21 \qquad b_{22} = (-3)(-4) + 5 \times 5 = 37$$

and

$$\mathbf{A}^2 = \begin{bmatrix} 16 & -28 \\ -21 & 37 \end{bmatrix}$$

Continuing in this manner, we obtain

$$\mathbf{A}^3 = \begin{bmatrix} 116 & -204 \\ -153 & 269 \end{bmatrix}$$

$$\mathbf{A}^5 = \mathbf{A}^2\mathbf{A}^3 = \mathbf{A}^3\mathbf{A}^2 = \begin{bmatrix} 6{,}140 & -10{,}796 \\ -8{,}097 & 14{,}237 \end{bmatrix}$$

If Eq. (3.8) is extended to include the case where $m = 0$, it becomes $\mathbf{A}^0\mathbf{A}^n = \mathbf{A}^n$. However, $\mathbf{I}\mathbf{A}^n = \mathbf{A}^n$, and it follows that

$$\mathbf{A}^0 = \mathbf{I} \tag{3.9}$$

3.7 DETERMINANTS

Associated with every square matrix is a numerical value called its *determinant*. The determinant of a matrix **A** is denoted by $|\mathbf{A}|$. Where reference is made to the determinant of a matrix rather than the matrix itself, the array of numbers is enclosed within vertical lines. If the determinant of a matrix is zero, the matrix is described as *singular*; if the determinant has a nonzero value, the matrix is described as *regular* or *nonsingular*.

We shall not define the determinant, but we shall present procedures for obtaining its value.

We start with the second-order matrix

$$\mathbf{A} = \begin{bmatrix} a_{11} & a_{12} \\ a_{21} & a_{22} \end{bmatrix}$$

Its determinant is

$$|\mathbf{A}| = a_{11}a_{22} - a_{12}a_{21} \tag{3.10}$$

EXAMPLE 3.10 Find the determinant of the following matrix:

$$\mathbf{A} = \begin{bmatrix} -17 & 13 \\ -8 & 21 \end{bmatrix}$$

SOLUTION

$$|\mathbf{A}| = (-17)21 - 13(-8) = -253$$

The determinant of a third-order matrix can be obtained by this device: Expand the matrix by repeating the first and second columns, keeping them in the same relative order and placing them to the right of the original matrix. Take each diagonal line in the expanded matrix on which three elements are located. Form the product of these elements. Consider the product positive if the diagonal inclines downward to the right and negative if it inclines downward to the left. The determinant equals the algebraic sum of these products.

EXAMPLE 3.11 Compute the determinant of the following matrix:

$$\mathbf{A} = \begin{bmatrix} 12 & 6 & -9 \\ 3 & 8 & 15 \\ 4 & 11 & 5 \end{bmatrix}$$

SOLUTION The expanded matrix is

$$\begin{bmatrix} 12 & 6 & -9 & 12 & 6 \\ 3 & 8 & 15 & 3 & 8 \\ 4 & 11 & 5 & 4 & 11 \end{bmatrix}$$

Proceeding in the prescribed manner, we obtain

$$|\mathbf{A}| = 12 \times 8 \times 5 + 6 \times 15 \times 4 + (-9)3 \times 11 - (-9)8 \times 4$$
$$- 12 \times 15 \times 11 - 6 \times 3 \times 5 = -1239$$

In Art. 3.9, we shall present a method of finding the determinant of a matrix of any order whatever.

3.8 MINORS AND COFACTORS

Let **A** denote a square matrix of order n. Now consider that the ith row and jth column of **A** are deleted, and let $\mathbf{S}_{ij}$ denote the submatrix that remains. Then $\mathbf{S}_{ij}$ is of the $(n-1)$th order. The determinant of $\mathbf{S}_{ij}$ is termed the *minor* of element a_{ij}, and it is denoted by $|\mathbf{M}_{ij}|$. The *cofactor* of element a_{ij}, which is denoted by $|\mathbf{A}_{ij}|$, is defined in this manner:

$$|\mathbf{A}_{ij}| = (-1)^{i+j}|\mathbf{M}_{ij}| \tag{3.11}$$

Thus, the cofactor and minor are coincident if $i+j$ is even, and they are of opposite algebraic sign if $i+j$ is odd. Figure 3.1 is presented as an aid in visualizing these definitions.

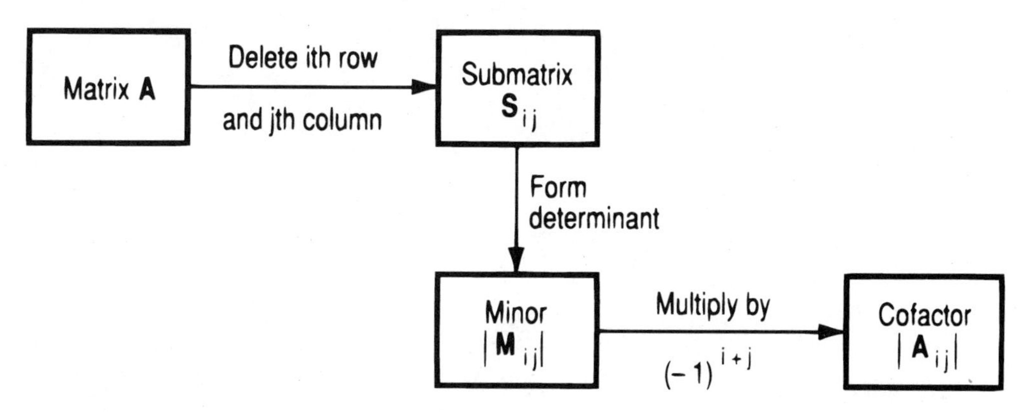

FIGURE 3.1 Method of forming cofactors.

EXAMPLE 3.12 With reference to the matrix in Example 3.11, find the cofactor of element a_{23}.

SOLUTION With the second row and third column eliminated, the submatrix that remains is

$$\mathbf{S}_{23} = \begin{bmatrix} 12 & 6 \\ 4 & 11 \end{bmatrix}$$

The determinant of this submatrix is the minor of a_{23}. By Eq. (3.10), its value is

$$|\mathbf{M}_{23}| = 12 \times 11 - 6 \times 4 = 108$$

The cofactor of a_{23} is

$$|\mathbf{A}_{23}| = -|\mathbf{M}_{23}| = -108$$

3.9 EVALUATION OF DETERMINANTS BY COFACTORS

The following principle affords a means of evaluating a determinant of any order:

Theorem 3.1. If all elements in a given row or column of a square matrix are multiplied by their cofactors, the sum of these products is equal to the determinant of the matrix.

EXAMPLE 3.13 Solve Example 3.11 by applying Theorem 3.1.

SOLUTION For convenience, we repeat the matrix.

$$\mathbf{A} = \begin{bmatrix} 12 & 6 & -9 \\ 3 & 8 & 15 \\ 4 & 11 & 5 \end{bmatrix}$$

We shall compute the cofactors of the elements in the third row. We apply Eq. (3.10), and the cofactors are as follows:

$$|\mathbf{A}_{31}| = \begin{vmatrix} 6 & -9 \\ 8 & 15 \end{vmatrix} = 6 \times 15 - (-9)8 = 162$$

$$|\mathbf{A}_{32}| = -\begin{vmatrix} 12 & -9 \\ 3 & 15 \end{vmatrix} = -[12 \times 15 - (-9)3] = -207$$

$$|\mathbf{A}_{33}| = \begin{vmatrix} 12 & 6 \\ 3 & 8 \end{vmatrix} = 12 \times 8 - 6 \times 3 = 78$$

$$|\mathbf{A}| = 4 \times 162 + 11(-207) + 5 \times 78 = -1239$$

This result coincides with that in Example 3.11.

The cofactor method of evaluating a determinant may be applied to determinants of higher order. However, it requires a cyclic procedure in which cofactors are successively reduced to lower order.

Another method of finding the determinant of a matrix is called *triangularization.* Under this method, the given matrix is transformed to a triangular matrix having the same determinant, and the determinant is then obtained by applying this principle: The determinant of a triangular matrix is equal to the product of the diagonal elements. A study of this method lies beyond the scope of this book.

3.10 INVERSE MATRIX

In Art. 3.7, we stated that a square matrix is regular if its determinant has a nonzero value. Associated with a regular matrix $\mathbf{A}$ is an *inverse* or *reciprocal matrix* $\mathbf{A}^{-1}$ such that

$$\mathbf{A}\mathbf{A}^{-1} = \mathbf{A}^{-1}\mathbf{A} = \mathbf{I}$$

where $\mathbf{I}$ is the unit matrix of the same order as $\mathbf{A}$.

Several methods of inverting a matrix are available. One method consists of the following steps:

1. Compute the determinant $|\mathbf{A}|$.
2. Replace $\mathbf{A}$ with its transpose $\mathbf{A}^T$.
3. Replace each element in $\mathbf{A}^T$ with its cofactor. The matrix that results is called the *adjoint* of $\mathbf{A}$, and it is denoted by adj $\mathbf{A}$.
4. Divide the adjoint of $\mathbf{A}$ by $|\mathbf{A}|$. The matrix thus formed is the inverse of $\mathbf{A}$.

Expressed symbolically,

$$\mathbf{A}^{-1} = \frac{\text{adj } \mathbf{A}}{|\mathbf{A}|} \tag{3.12}$$

In obtaining the adjoint of $\mathbf{A}$, we may take steps 2 and 3 in reverse order; i.e., we may replace $\mathbf{A}$ with its cofactors, and then replace the resulting matrix with its transpose.

EXAMPLE 3.14 Construct the inverse of the matrix in Example 3.11.

SOLUTION The transpose matrix, which we shall call $\mathbf{B}$, is

$$\mathbf{B} = \mathbf{A}^T = \begin{bmatrix} 12 & 3 & 4 \\ 6 & 8 & 11 \\ -9 & 15 & 5 \end{bmatrix}$$

In finding the cofactors of $\mathbf{B}$, we recommend that the reader repeat the matrix in each instance and then cross out the appropriate row and column. The cofactors are as follows:

$$|\mathbf{B}_{11}| = 8 \times 5 - 11 \times 15 = -125$$

$$|\mathbf{B}_{12}| = -[6 \times 5 - 11(-9)] = -129$$

$$|\mathbf{B}_{13}| = 6 \times 15 - 8(-9) = 162$$

$$|\mathbf{B}_{21}| = -(3 \times 5 - 4 \times 15) = 45$$

$$|\mathbf{B}_{22}| = 12 \times 5 - 4(-9) = 96$$

$$|\mathbf{B}_{23}| = -[12 \times 15 - 3(-9)] = -207$$

$$|\mathbf{B}_{31}| = 3 \times 11 - 4 \times 8 = 1$$

$$|\mathbf{B}_{32}| = -(12 \times 11 - 4 \times 6) = -108$$

$$|\mathbf{B}_{33}| = 12 \times 8 - 3 \times 6 = 78$$

In Example 3.11, we found that $|\mathbf{A}| = -1239$. Changing the algebraic sign of each element, we obtain

$$\mathbf{A}^{-1} = \frac{1}{1239} \begin{bmatrix} 125 & 129 & -162 \\ -45 & -96 & 207 \\ -1 & 108 & -78 \end{bmatrix}$$

The result can be verified by performing the multiplication $\mathbf{AA}^{-1}$. We find that the product is the unit matrix, and it follows that our inverse matrix is correct.

CHAPTER 4

STATISTICS OF DISCRETE VARIABLES

4.1 DEFINITIONS

Consider that a variable X can assume different values on different occasions. The set of values assumed by X is termed *statistical data*, and the value assumed by X on a specific occasion is termed an *element*. We shall confine our investigation at present to discrete variables.

Let n denote the number of times that X assumed a value, and let f_i denote the number of times that X assumed the value X_i. The quantity f_i is called the *frequency* of X_i, and n equals the sum of the frequencies. The quantity f_i/n is termed the *relative* frequency of X_i. Relative frequency can be expressed in the form of an ordinary fraction, a decimal fraction, or a percent. The sum of the relative frequencies of all values of X is perforce 1.

A *cumulative frequency* is the sum of the frequencies of X up to or beyond a certain value; it may or may not include that value. Cumulative frequency can also be expressed on a relative basis.

The values of X and their respective frequencies (or relative frequencies) constitute the *frequency distribution* of X.

EXAMPLE 4.1 Table 4.1 presents the frequency distribution of a variable X. Compute the following cumulative frequencies: the number of times X was less than 7; the number of times X was at least 7; the number of times X was more than 7.

SOLUTION The number of times X was less than 7 is $5 + 9 + 17 = 31$. The

TABLE 4.1

Value of X	Frequency
4	5
5	9
6	17
7	13
8	4
9	3
Total	51

number of times X was at least 7 is $13 + 4 + 3 = 20$. The number of times X was more than 7 is $4 + 3 = 7$.

EXAMPLE 4.2 With reference to Table 4.1, compute the relative number of times that X was less than 8 (in percentage form).

SOLUTION The number of times that X was less than 8 is $5 + 9 + 17 + 13 = 44$ (or $51 - 4 - 3 = 44$). The relative number of times that X was less than 8 is $44/51 = 86.27$ percent.

4.2 GROUPING OF DATA

Where the number of values of X is very large, the statistical data are presented by grouping the values of X in *classes* (or *cells*) and recording the frequency of each class. Table 4.2 illustrates the method of grouping for a variable that is restricted to integral values. The range of values associated with a given class is called the *class interval*, and the end values of the class interval are called the *class limits*.

TABLE 4.2

Class interval	Frequency
101 to 105	6
106 to 110	13
111 to 115	22
116 to 120	19
121 to 125	11
126 to 130	5
131 to 135	2
Total	78

For analytic purposes, it is essential that there be no gaps between the classes. Therefore, although realistically the variable in Table 4.2 can assume only integral values, it is necessary nevertheless to extend each class interval one-half unit on each side to secure continuity. For example, the true interval for the second class is 105.5 to 110.5. The difference between the upper and lower limits of a class is known as the *class width*.

By convention, it is assumed that the values that lie within a given class are uniformly distributed through the class interval. Thus, each interval is divided into uniform *subintervals*. Moreover, the values are assumed to lie at the *upper ends* of the subintervals.

EXAMPLE 4.3 With reference to Table 4.2, determine the seventh value in the fourth class by applying the conventional method.

SOLUTION The true class interval is 115.5 to 120.5, and the class width is 5. The interval is divided into 19 uniform subintervals, and the seventh value lies at the upper end of the seventh subinterval. Therefore, the seventh value in this class is $115.5 + (7/19)5 = 117.34$ to two decimal places.

The *midpoint* or *mark* of a class is the value of the variable that lies midway between the upper and lower limits. For example, the midpoint of the third class in Table 4.2 is 113.

In the subsequent material, it is understood that all classes are of uniform width.

4.3 FREQUENCY-DISTRIBUTION DIAGRAMS

Statistical data can be visualized more readily if we construct a diagram that exhibits the frequency distribution. In this diagram, values of the variable are plotted on the horizontal axis.

For ungrouped data, the diagram has the simple form illustrated in Fig. 4.1, which is a plot of the data in Table 4.1. A vertical line is drawn above each value of X, the length of the line being equal to the frequency of that value on the basis of some suitable scale.

Grouped data can be represented by a *histogram*, which is a diagram composed of rectangles. A rectangle is constructed above each class interval, the area of the rectangle being equal to the frequency of that class. If the classes are of uniform width, the height of each rectangle is directly proportional to the frequency of that class. Thus, frequencies (or relative frequencies) can be plotted on the vertical axis. A histogram is illustrated by Fig. 4.2, which is a plot of the data in Table 4.2.

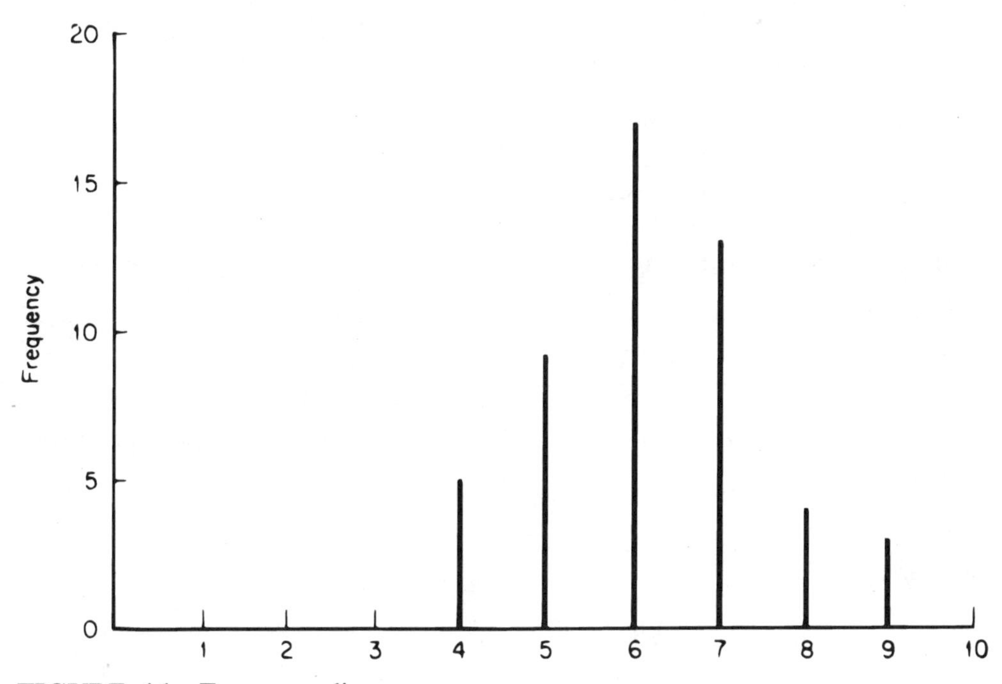

FIGURE 4.1 Frequency diagram.

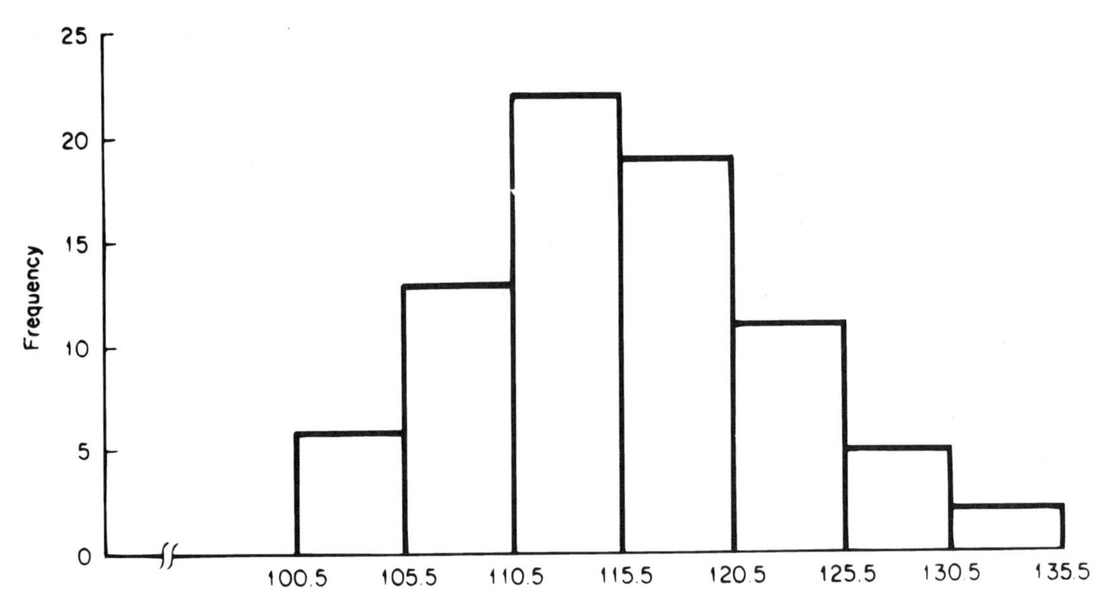

FIGURE 4.2 Histogram.

4.4 DEFINITION OF AVERAGE AND MEAN

In statistical analysis, it is advantageous to replace all values assumed by X with some constant value that is representative of the entire set. This representative value is termed an *average*, and the most frequently applied type of average is the mean. We shall first present a general definition of the mean.

The set of values assumed by X has certain *properties*: the sum of the X values, the sum of the X^2 values, the product of the X values, etc. Let X_m denote a constant value that can replace each X value without changing some specific property. This quantity X_m is called a *mean* of the X values. For example, consider that X assumed the values 12, 15, and 21 on three separate occasions. The sum of the elements is 48, and it is possible to replace each X value with 16 without changing the sum. Therefore, 16 is a mean of the given set of X values. Similarly, the product of the elements is 3780, and it is possible to replace each X value with 15.5775 without changing the product. Therefore, 15.5775 is also a mean of the given set of X values.

Since a set of X values has numerous properties, it follows that it also has numerous means, each corresponding to a specific property. We shall define and investigate the means that are of greatest significance, and we shall assign a distinctive symbol to each. As before, we shall let n denote the sum of the frequencies.

4.5 ARITHMETIC MEAN

The arithmetic mean is the value of X that would produce the same *sum* as the true X values. It is denoted by $\bar{X}$.

EXAMPLE 4.4 Find the arithmetic mean of the set of values in Table 4.1 (to four decimal places).

SOLUTION The sum of the X values is

$$5 \times 4 + 9 \times 5 + 17 \times 6 + 13 \times 7 + 4 \times 8 + 3 \times 9 = 317$$

Since X assumed a value on 51 occasions, we have

$$51\bar{X} = 317 \qquad \bar{X} = 6.2157$$

In general,

$$\bar{X} = \frac{\sum fX}{n} \tag{4.1}$$

This equation is also applicable to grouped data if we set all X values within a class equal to the midpoint of that class.

EXAMPLE 4.5 Table 4.3 presents the income earned by members of an organization during a given period of time. Compute the arithmetic mean of the income (to three significant figures).

TABLE 4.3

Income, \$ (class interval)	Number of members (frequency)
24,000 to less than 26,000	2
26,000 to less than 28,000	5
28,000 to less than 30,000	9
30,000 to less than 32,000	6
32,000 to less than 34,000	4
34,000 to less than 36,000	1
Total	27

SOLUTION Applying the class midpoints, we obtain the following:

$$\begin{aligned}\sum fX &= 2 \times 25{,}000 + 5 \times 27{,}000 + 9 \times 29{,}000 + 6 \times 31{,}000 + 4 \times 33{,}000 + 35{,}000 \\ &= \$799{,}000\end{aligned}$$

$$\bar{X} = \frac{799{,}000}{27} = \$29{,}600$$

As before, let X_i denote a specific value of X and let $d_{m,i} = X_i - \bar{X}$. The quantity $d_{m,i}$ is termed the *deviation* of X_i from $\bar{X}$. Equation (4.1) yields this relationship:

$$\sum fd_m = 0 \tag{4.2}$$

Expressed verbally, the algebraic sum of the deviations from the arithmetic mean is zero.

4.6 WEIGHTED ARITHMETIC MEAN

It is often necessary to compute the arithmetic mean of a set of X values while recognizing that these values are of varying importance. In a situation of this type, we assign *weights* to the X values on the basis of their relative importance and then compute a *weighted* arithmetic mean of these values.

Let w_i denote the weight assigned to the value X_i, and let $\bar{X}_w$ denote the weighted arithmetic mean. Then

$$\bar{X}_w = \frac{\sum wfX}{\sum wf} \tag{4.3}$$

EXAMPLE 4.6 A student received the examination grades recorded in column 2 of Table 4.4. To arrive at the term grade, the instructor assigned to the examinations the weights shown in column 3. Establish the term grade of this student.

TABLE 4.4

Examination number (1)	Grade (2)	Weight (3)	Product (4)
1	61	1.00	61.00
2	75	1.50	112.50
3	83	1.75	145.25
4	85	2.00	170.00
Total		6.25	488.75

SOLUTION Multiply the grades by their respective weights to obtain the products shown in column 4. Total columns 3 and 4. Then

$$\text{Term grade} = \frac{488.75}{6.25} = 78.2$$

Weighted arithmetic means abound in engineering. For example, the specific heat of a mixture of perfect gases is the weighted arithmetic mean of the specific heats of the component gases, the statistical weights being the masses of the component gases.

4.7 THE MEDIAN AND THE MODE

As a representative value of a set of data, the arithmetic mean has this serious weakness: It is strongly influenced by an extreme value and thus presents a distorted view of a set of data. Therefore, it is advantageous to apply some other form of average to be used in conjunction with the arithmetic mean.

One form of average that is often applied is the *median*. Consider that all elements in the set of X values are arranged in ascending order of magnitude. This arrangement is called the *ascending array* of the elements. If the number of elements n is odd, the median is the value that occupies the central position in this array. [The position number is $(n + 1)/2$.] If the number of elements is even, the median is equated to the arithmetic mean of the two values that lie at the center of the array. (Their position numbers are $n/2$ and $n/2 + 1$.)

EXAMPLE 4.7 Find the median of the following set of X values:

Value	12	13	15	16	21	32
Frequency	1	4	2	2	1	1

SOLUTION The ascending array is as follows:

12 13 13 13 13 15 15 16 16 21 32

Since $n = 11$, the position number at the center is $12/2 = 6$. Therefore, the median is equal to the sixth element in this array, or 15. (The arithmetic mean is 16.27.)

EXAMPLE 4.8 Find the median of the following set of X values:

Value	5	7	22	23	24	25	26
Frequency	1	1	8	7	10	5	2

SOLUTION Since $n = 34$, the position numbers at the center are $34/2 = 17$ and 18. By taking cumulative frequencies up to a given value, we find that the seventeenth element is 23 and the eighteenth element is 24. Therefore, the median is 23.5. (The arithmetic mean is 22.53.)

Another form of average that is often applied in conjunction with the arithmetic mean is the *mode*. The mode is simply the X value that has the highest frequency.

EXAMPLE 4.9 Find the mode of the following set of X values:

Value	7	8	9	10	11	12
Frequency	2	3	12	15	9	5

SOLUTION The mode is 10.

EXAMPLE 4.10 Find the mode of the following set of X values:

Value	10	11	12	13	14	15
Frequency	2	6	9	9	7	4

SOLUTION This set of values has *two* modes: 12 and 13. The set is described as *bimodal.*

4.8 GEOMETRIC MEAN

The geometric mean is the value of X that would produce the same *product* as the true X values. It is denoted by X_G.

EXAMPLE 4.11 Find the geometric mean of the following set of X values:

Value	5	6	7	8
Frequency	1	2	3	2

SOLUTION The product of the X values is

$$5 \times 6^2 \times 7^3 \times 8^2 = 3{,}951{,}360$$

Since $n = 8$, we have

$$X_G^8 = 3{,}951{,}360 \qquad X_G = 3{,}951{,}360^{1/8} = 6.6772$$

Let $X_1, X_2, \ldots, X_k$ denote the values assumed by X and $f1, f2, \ldots, fk$ denote their respective frequencies. Then

$$X_G = (X_1^{f1} X_2^{f2} \cdots X_k^{fk})^{1/n} \tag{4.4}$$

4.9 HARMONIC MEAN

The harmonic mean is the value of X that would cause the *sum of the reciprocals of* X to remain constant. It is denoted by X_H.

EXAMPLE 4.12 Find the harmonic mean of the following set of X values:

Value	4	5	8	10
Frequency	2	2	3	1

SOLUTION The sum of the reciprocals of the X values is

$$\frac{2}{4} + \frac{2}{5} + \frac{3}{8} + \frac{1}{10} = 0.500 + 0.400 + 0.375 + 0.100 = 1.375$$

Since $n = 8$, we have

$$\frac{8}{X_H} = 1.375 \qquad X_H = \frac{8}{1.375} = 5.8182$$

In general,

$$X_H = \frac{n}{\sum (f/X)} \tag{4.5}$$

EXAMPLE 4.13 Points A, B, C, and D are equal distances apart. A vehicle traveled from A to D at the following mean speeds: from A to B, 105 km/h; from B to C, 92 km/h; from C to D, 83 km/h. Compute the mean speed S_m of the vehicle in traveling from A to D (i.e., the speed at which the total time of travel would be the same). Verify the result.

SOLUTION Since time equals distance divided by speed, and distance is constant, the mean speed from A to D is the harmonic mean of the three specified speeds.

$$\frac{1}{105} + \frac{1}{92} + \frac{1}{83} = 0.0324416$$

$$S_m = \frac{3}{0.0324416} = 92.4739 \text{ km/h}$$

Verification. Arbitrarily assume that the distance between two successive points is 300 km. The time of travel is 2.8571 h from A to B, 3.2609 h from B to C, and 3.6145 h from C to D, and the total time of travel from A to D is 9.7325 h. If the vehicle had traveled at the constant speed of 92.4739 km/h, the time of travel from A to D would have been

$$\frac{3 \times 300}{92.4739} = 9.7325 \text{ h}$$

Our calculated value of S_m is thus confirmed.

4.10 ROOT MEAN SQUARE

The root mean square (or quadratic mean) is the value of X that would produce the same *sum of the X^2 values.* It is denoted by X_{rms}.

EXAMPLE 4.14 Find the root mean square of the X values in Example 4.12.

SOLUTION The sum of the X^2 values is

$$2 \times 4^2 + 2 \times 5^2 + 3 \times 8^2 + 10^2 = 374$$

Since $n = 8$, we have

$$8X_{\text{rms}}^2 = 374 \qquad X_{\text{rms}} = \sqrt{\frac{374}{8}} = 6.8374$$

In general,

$$X_{\text{rms}} = \sqrt{\frac{\sum fX^2}{n}} \tag{4.6}$$

Thus, the expression *root mean square* is a contraction of the expression *square root of the arithmetic mean of the squares.*

The root mean square is applied extensively in electrical engineering, where we calculate the rms values of a voltage and current that vary with time.

4.11 STANDARD DEVIATION AND VARIANCE

When a set of X values is given, it is often essential to know whether these values cluster within a narrow range or are widely scattered. Thus, the *dispersion* of the X values is highly significant.

In Art. 4.5, we defined the deviation of an X value from the arithmetic mean, and the use of an average value of these deviations suggests itself as an effective means of gaging dispersion. However, in accordance with Eq. (4.2), the arithmetic mean of the deviations is zero. To overcome this obstacle, we *square* the deviations to make them all positive, take the arithmetic mean of the squared deviations, and then extract the square root of the value thus obtained. The result is called the *standard deviation* of the set of X values, and it is denoted by s. Then

$$s = \sqrt{\frac{\sum fd_m^2}{n}} = \sqrt{\frac{\sum f(X - \bar{X})^2}{n}} \tag{4.7}$$

A comparison of Eqs. (4.6) and (4.7) reveals that the standard deviation is the root mean square of the individual deviations. The quantity s^2 is termed the *variance* of the set of X values.

EXAMPLE 4.15 Tables 4.5 and 4.6 each exhibit, in columns 1 and 2, the values assumed by X and their respective frequencies. Compute the standard deviation of each set of values.

SOLUTION In each table, compute the values shown in column 3 and obtain the total of columns 2 and 3. In both instances, $\sum f = 20$ and $\sum fX = 426$. Therefore,

$$\bar{X} = \frac{426}{20} = 21.3$$

Now compute the deviations from the arithmetic mean; they are recorded in column 4. Compute the values shown in column 5 and obtain their total. For the set

TABLE 4.5

X (1)	Frequency, f (2)	fX (3)	$d_m = X - 21.3$ (4)	fd_m^2 (5)
20	3	60	−1.3	5.07
21	9	189	−0.3	0.81
22	7	154	0.7	3.43
23	1	23	1.7	2.89
Total	20	426		12.20

TABLE 4.6

X (1)	Frequency, f (2)	fX (3)	$d_m = X - 21.3$ (4)	fd_m^2 (5)
10	2	20	−11.3	255.38
13	1	13	−8.3	68.89
17	6	102	−4.3	110.94
20	5	100	−1.3	8.45
29	3	87	7.7	177.87
34	2	68	12.7	322.58
36	1	36	14.7	216.09
Total	20	426		1160.20

of values in Table 4.5,

$$s = \sqrt{\frac{12.20}{20}} = 0.7810$$

For the set of values in Table 4.6,

$$s = \sqrt{\frac{1160.20}{20}} = 7.6164$$

A cursory examination of the values in Tables 4.5 and 4.6 makes it apparent that the set in the second table is far more widely scattered than that in the first table. Our calculations provide a quantitative comparison.

4.12 STANDARDIZED VARIABLES AND STANDARD UNITS

The deviation of a given value X_i from the arithmetic mean $\bar{X}$ must often be expressed on a *relative* basis for the result to be meaningful. A relative value can be obtained by dividing the deviation by the standard deviation, and the result is termed a *standardized variable*. Let z_i denote this quantity. Then

$$z_i = \frac{d_{m,i}}{s} = \frac{X_i - \bar{X}}{s} \tag{4.8}$$

The quantity z_i, which is a pure number, represents the number of standard deviations contained in the given deviation. Therefore, it is said to be expressed in *standard units*.

EXAMPLE 4.16 A class was given an examination in astronomy and in chemistry. In the astronomy examination, the mean grade was 66 and the standard deviation was 3. In the chemistry examination, the mean grade was 78 and the standard deviation was 8. A student received a grade of 70 in astronomy and 85 in chemistry. Evaluate the relative superiority of this student in each examination.

SOLUTION The relative superiority is as follows:

Astronomy: $$z = \frac{70 - 66}{3} = 1.333$$

Chemistry: $$z = \frac{85 - 78}{8} = 0.875$$

Thus, although this student's grade was 4 points above average in the astronomy examination and 7 points above average in the chemistry examination, the student fared better in astronomy than in chemistry. This paradox is explained by the fact that the grades were more widely scattered in chemistry than in astronomy.

CHAPTER 5

PROBABILITY OF DISCRETE VARIABLES

5.1 DEFINITIONS

If the value that a variable will assume on a given occasion cannot be predetermined because it is influenced by chance, this quantity is referred to as a *random* or *stochastic variable*. For example, the number of defective parts that will be present in the next shipment of parts received by a business firm is a random variable. At present, we shall confine our investigation to *discrete* random variables.

A process that yields a value of the random variable is called a *trial* or *experiment*, and the value the variable assumes in a given trial is called the *outcome*. As an illustration, assume that a bowl contains nine chips that are numbered consecutively from 1 to 9 and that a chip will be drawn at random from the bowl. Let X denote the number on the chip that is drawn. Then X is a random variable, the process of drawing a chip is the trial, and the value of X is the outcome. There are nine possible outcomes.

In the study of probability, the term *variable* is applied in a broad sense that extends beyond numerical quantities. As an illustration, assume the following: A bowl contains two chips; one bears the letter A and the other bears the letter B; a chip will be drawn at random. Since the letter on the chip that is drawn will be A or B, this letter is a variable.

A specified outcome or set of outcomes is termed an *event*. For example, with reference to drawing a chip, we may define the following events: The outcome is 4; the outcome is even but less than 8 (which comprises the outcomes 2, 4, and 6); the outcome is at least 8 (which comprises the outcomes 8 and 9). The letter E is generally used to denote an event.

Two events are *mutually exclusive* or *disjoint* if either one of them, but not both, can result from a single trial. For example, with reference to drawing a chip, the following events are mutually exclusive: The outcome is even; the outcome is 5, 7, or 9.

Two events are *overlapping* if there is at least one outcome that will satisfy both events. For example, with reference to drawing a chip, consider the following pair of events: The outcome is even; the outcome is less than 6. There are two outcomes (2 and 4) that satisfy both events. Therefore, these events are overlapping.

A set of events is *exhaustive* if it includes all possible outcomes. Thus, with reference to drawing a chip, the following set of events is exhaustive: The outcome is even; the outcome is odd and less than 9; the outcome is 9.

Associated with every event E is another event $\bar{E}$ that is the negation of E. For example, if E denotes that a switch is closed, $\bar{E}$ denotes that it is open. Similarly, if E denotes that a machine part is satisfactory, $\bar{E}$ denotes that it is defective. Events E and $\bar{E}$ are said to be *complementary* to each other.

Assume that a trial has n possible outcomes and that one outcome is just as likely as any other. The *probability* of a particular outcome is defined as $1/n$.

5.2 VENN DIAGRAMS

The probability of a given event can be visualized more readily by constructing a *Venn diagram.* In this diagram, each possible outcome of a trial is represented by a unique point on a plane, and it is termed a *sample point*. The region of the plane occupied by the sample points corresponding to a given trial is termed the *sample space* of the trial.

With reference to Fig. 5.1*a*, consider that the sample points corresponding to a given trial lie within the rectangle *abcd*. This rectangle is the sample space of the trial. Assume that an event E is satisfied solely by outcomes having sample points within the shaded circle. This circle *represents* event E. The events E_1 and E_2 represented by the shaded circles in Fig. 5.1*b* are mutually exclusive because they have no outcomes in common. On the other hand, the events E_3 and E_4 represented by the shaded circles in Fig. 5.1*c* are overlapping because they do have such outcomes. It is apparent that the Venn diagrams of probability are completely analogous to the Venn diagrams of set theory presented in Chap. 2.

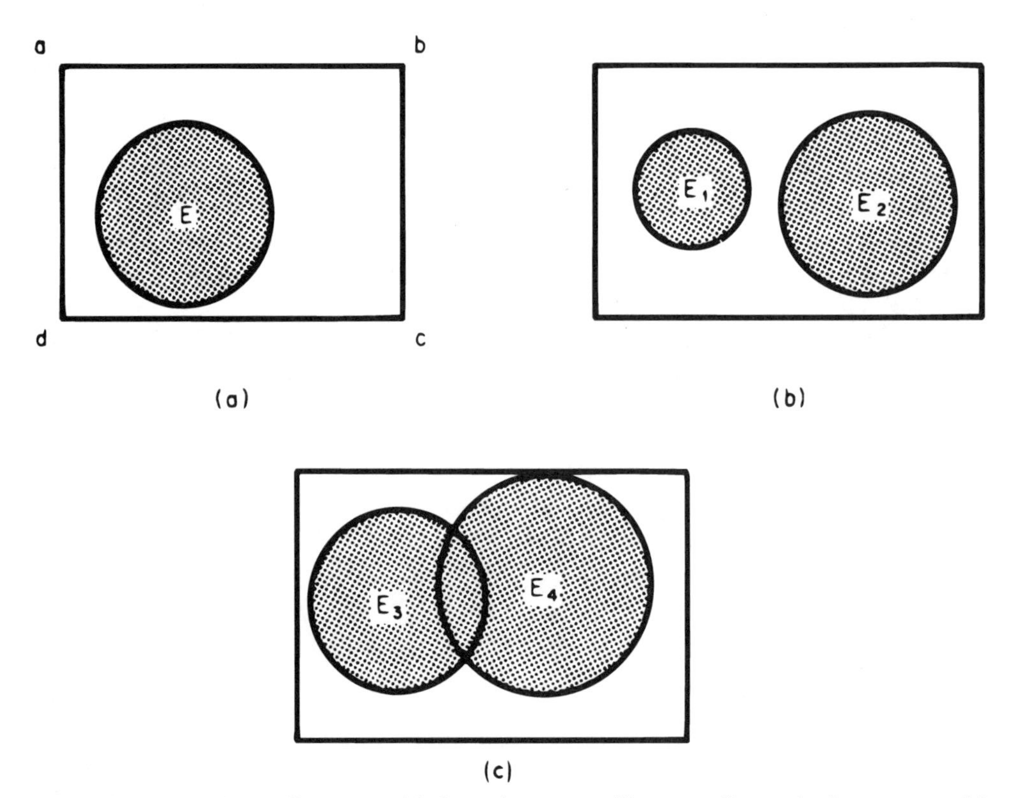

FIGURE 5.1 Venn diagrams. (*a*) Sample space; (*b*) mutually exclusive events; (*c*) overlapping events.

5.3 LAWS OF PROBABILITY

There are several simple laws that underlie the calculation of probability, and we shall present four of these laws at this point. The first law pertains to a single event, the second to mutually exclusive events, the third to independent events, and the fourth to overlapping events. In Art 5.4, we shall present the law that pertains to dependent events.

The probability of an event E is denoted by $P(E)$.

Theorem 5.1. Assume that a trial has n possible outcomes of equal probability. If any one of r outcomes will produce an event E, the probability that E will occur is $P(E) = r/n$.

EXAMPLE 5.1 With reference to the bowl of chips described in Art. 5.1, assume that all chips have the same likelihood of being drawn. What is the probability that X will be even?

SOLUTION This event is satisfied by four of the nine possible outcomes (namely, 2, 4, 6, and 8). Therefore, the probability of this event is 4/9.

It follows as a corollary of Theorem 5.1 that an impossible event has a probability of 0 and an event that is certain to occur has a probability of 1. Therefore, the probability of an event can range from 0 to 1, inclusive. The value can be expressed as an ordinary fraction, a decimal fraction, or a percent.

There are many situations where a permutation or combination is to be constructed at random, and it becomes necessary to find the total number of permutations or combinations that can be constructed in order to calculate probability. The equations of Arts. 1.23 and 1.24 are applicable. It is understood that all permutations or combinations have equal likelihood of being the one that is constructed.

EXAMPLE 5.2 A four-place number will be formed by selecting integers from 1 to 7, inclusive, at random. Each digit can appear only once in the number. What is the probability that the number formed will contain the digits 2, 3, and 7, without reference to their order?

SOLUTION Since we are concerned solely with the identity of the digits that form the number, each selection of digits is a *combination* of seven digits taken four at a time. The number of possible combinations is $C(7,4) = 7!/(4!\,3!) = 35$. The digits 2, 3, and 7 can be combined with any of the remaining four digits; therefore, the number of satisfactory combinations is 4. Thus, the probability of the specified event is $4/35 = 0.1143$.

EXAMPLE 5.3 A bowl contains 10 chips that are numbered consecutively from 1 to 10. Six chips will be drawn at random from the bowl and placed in a row. What is the probability that the chips marked 2, 3, and 7 will occupy, respectively, the first, fourth, and fifth positions in the row?

SOLUTION The number of ways in which the row can be formed is $P(10,6) = 10!/4! = 151{,}200$. Consider that the chips marked 2, 3, and 7 are placed in the designated positions. Three positions remain to be filled, and seven chips are now available. Therefore, the number of ways in which the row can be completed is

$P(7,3) = 7!/4! = 210$. Thus, the probability of the specified event is 210/151,200 = 0.001389.

EXAMPLE 5.4 A committee consisting of six members of equal rank is to be formed. Fifteen individuals, including Smith and Jones, are available for appointment to the committee. The committee members will be selected randomly. What is the probability that the committee will include Jones but not Smith?

SOLUTION Since the committee members will be of equal rank, each possible committee represents a *combination* of 15 individuals taken 6 at a time. The number of possible committees if no restrictions are imposed is $C(15,6) = 15!/(6!9!) = 5005$. Consider that Jones has been appointed to the committee but Smith is to be excluded. The five remaining committee members can be selected from 13 individuals, and the number of ways in which the committee can be completed is $C(13,5) = 13!/(5!8!) = 1287$. Therefore, the probability that the committee formed will meet the specifications is 1287/5005 = 0.2571.

Theorem 5.2. If two events E_1 and E_2 are mutually exclusive, the probability that either E_1 or E_2 will occur is the sum of their respective probabilities. Expressed symbolically,

$$P(E_1 \text{ or } E_2) = P(E_1) + P(E_2)$$

Figure 5.1*b* makes this relationship self-evident, and the principle can be extended to include any number of mutually exclusive events.

EXAMPLE 5.5 A case contains an assortment of machine parts, and a part will be drawn at random. The probability is 0.09 that the part that is drawn is type A, 0.32 that it is type B, and 0.21 that it is type C. Compute the probability that the part that is drawn is either type A or type B, either type A or type C, either type B or type C, or any of these three types.

SOLUTION We are concerned with three possible events: drawing type A, drawing type B, and drawing type C. These events are mutually exclusive. Then

$$P(\text{A or B}) = 0.09 + 0.32 = 0.41$$

$$P(\text{A or C}) = 0.09 + 0.21 = 0.30$$

$$P(\text{B or C}) = 0.32 + 0.21 = 0.53$$

$$P(\text{A, B, or C}) = 0.09 + 0.32 + 0.21 = 0.62$$

Since it is a certainty that a given event E either will or will not result from a trial, it is a certainty that either E or its complementary event $\bar{E}$ will occur. Since the probability of certainty is 1, it follows from Theorem 5.2 that the sum of their probabilities is 1, or

$$P(\bar{E}) = 1 - P(E) \tag{5.1}$$

For example, assume that an object will be drawn at random from a box. If the probability of drawing a yellow object is 0.47, the probability of drawing an object of some other color is $1 - 0.47 = 0.53$.

EXAMPLE 5.6 With reference to Example 5.5, what is the probability that the part that is drawn is neither type B nor type C?

SOLUTION Since the probability that this part is either type B or C is 0.53, the probability that it is neither of these types is 1 − 0.53 = 0.47.

Two trials are said to be *independent* of each other if the outcome of one trial has no bearing on the outcome of the other trial.

Theorem 5.3. Assume that two independent trials will be performed. The probability that the first trial will produce an event E_1 and the second trial will produce an event E_2 is the product of their respective probabilities. Expressed symbolically,

$$P(E_1 \text{ and } E_2) = P(E_1) \times P(E_2)$$

This statement can be extended to include any number of independent trials.

EXAMPLE 5.7 Three individuals, A, B, and C, are working independently on their respective assignments. The probability that an individual will complete an assignment on time is as follows: A, 0.76; B, 0.81; C, 0.88. What is the probability that all three individuals will complete their assignments on time?

SOLUTION Since the success or failure of one individual has no bearing on the success or failure of any other individual, the three results are independent of one another. By Theorem 5.3, the probability of the specified event is

$$(0.76)(0.81)(0.88) = 0.5417$$

EXAMPLE 5.8 With reference to Example 5.7, what is the probability that not all three individuals will complete their assignments on time?

SOLUTION The event now specified is the complement of that specified in Example 5.7. By Eq. (5.1), the probability that not all three individuals will complete their assignments on time is 1 − 0.5417 = 0.4583.

EXAMPLE 5.9 With reference to Example 5.7, what is the probability that A and C will complete their assignments on time but B will fail to do so?

SOLUTION The probability is

$$(0.76)(1 - 0.81)(0.88) = 0.1271$$

EXAMPLE 5.10 A building has three entrances: L, M, and N. The probability that an entrance is open at a given instant is as follows: L, 0.16; M, 0.09; N, 0.14. If an individual wishes to enter the building when it is vacant but has lost his key, what is the probability that he will be unable to enter?

SOLUTION This individual will be unable to enter the building if all three entrances are simultaneously closed. The probability of this condition is

$$(1 - 0.16)(1 - 0.09)(1 - 0.14) = 0.6574$$

EXAMPLE 5.11 Solve Example 5.10 with this modification: The building has a fourth entrance, and the probability that this entrance is open at a given instant is 0.17.

SOLUTION The probability that the individual will fail to gain entrance now becomes

$$(1 - 0.16)(1 - 0.09)(1 - 0.14)(1 - 0.17) = 0.5456$$

As the number of entrances increases, the probability that the individual *will* gain entrance increases.

Consider a situation of this type: We have a set of events that can stem from a series of independent trials, and an event E results from the occurrence of any event in this set, or any combination of these events. If the number of events in the set is more than two, the number of possible combinations is large. Therefore, $P(E)$ can be found most expeditiously by an indirect approach. We calculate the probability of the complementary event $\bar{E}$ and then subtract the result from 1, in accordance with Eq. (5.1). We shall illustrate the procedure.

EXAMPLE 5.12 At an amusement center, four individuals, A, B, C, and D, fire simultaneously at a moving target. The probability that an individual will strike the target is as follows: A, 0.35; B, 0.42; C, 0.44; D, 0.29. What is the probability that the target will be struck?

SOLUTION The target is struck if *at least one individual* is successful. The probability that the target *is not struck* is

$$(1 - 0.35)(1 - 0.42)(1 - 0.44)(1 - 0.29) = 0.1499$$

Therefore, the probability that the target *is struck* is $1 - 0.1499 = 0.8501$.

EXAMPLE 5.13 With reference to the electrical network in Fig. 5.2, all five relays function independently, and the probability that any relay is closed is p. What is the probability that a current exists between a and b?

SOLUTION There are three alternative paths from a to b, and these are assigned the numbers shown in circles. A current exists if all relays along *any* path are closed.

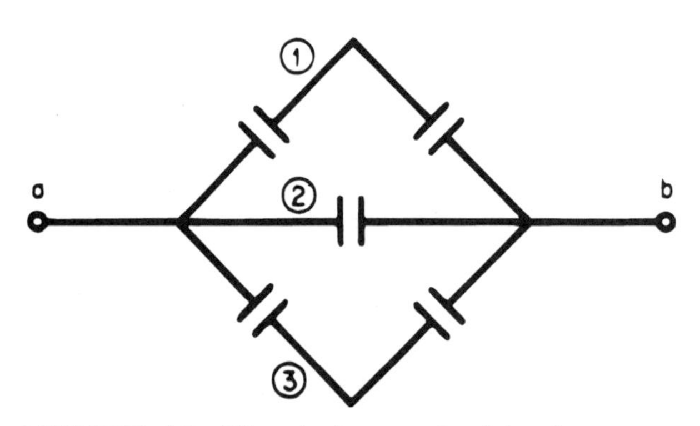

FIGURE 5.2 Electrical network with relays.

By Theorem 5.3, the probability of complete closure along a path is p^r, where r is the number of relays along the path. Therefore, the probability of complete closure along a path has the following values: path 1, p^2; path 2, p; path 3, p^2. The probability that there is no closure along any path is

$$(1 - p^2)(1 - p)(1 - p^2) = 1 - p - 2p^2 + 2p^3 + p^4 - p^5$$

and the probability that a current exists is

$$p + 2p^2 - 2p^3 - p^4 + p^5$$

EXAMPLE 5.14 Cases A, B, C, and D contain machine parts, and the incidence of defective parts is 3.2 percent in A, 2.7 percent in B, 1.9 percent in C, and 4.1 percent in D. To manufacture a product, it is necessary to draw a part from each case. If the parts are drawn at random, what is the probability that the set of parts that is drawn is unsatisfactory?

SOLUTION The set of parts is unsatisfactory if at least one part is defective. The probability that all four parts are *satisfactory* is

$$(1 - 0.032)(1 - 0.027)(1 - 0.019)(1 - 0.041) = 0.8861$$

Therefore, the probability that at least one part is defective is $1 - 0.8861 = 0.1139$ or 11.39 percent.

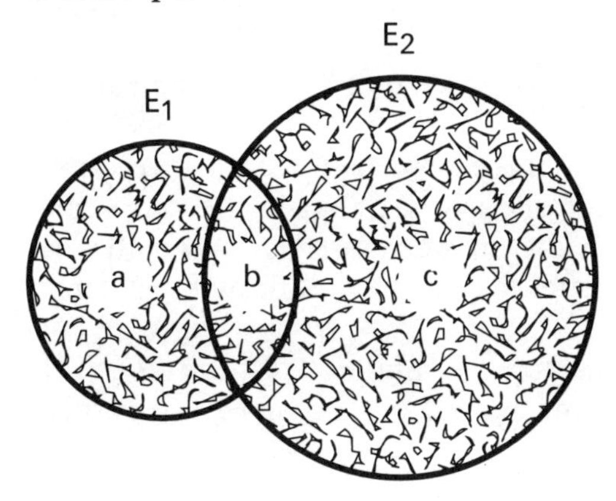

FIGURE 5.3 Venn diagram for two overlapping events.

Theorem 5.4. Let E_1 and E_2 denote two overlapping events. If an event E results from the occurrence of E_1 or E_2, or both, the probability of E is

$$P(E) = P(E_1) + P(E_2) - P(E_1 \text{ and } E_2)$$

The validity of this statement follows from the Venn diagram in Fig. 5.3, where

$$P(E) = a + b + c \qquad P(E_1) = a + b$$

$$P(E_2) = b + c \qquad P(E_1 \text{ and } E_2) = b$$

By drawing the corresponding Venn diagram, it becomes possible to express $P(E)$ where there are three or more overlapping events.

EXAMPLE 5.15 Three objects, A, B, and C, are placed in a row, in that order. The probability that an object is illuminated at a given instant is as follows: A, 0.62; B, 0.55; C, 0.71. It is necessary that at least two adjacent objects be illuminated simultaneously. What is the probability that the requirement is satisfied?

SOLUTION

Method 1. We define the following events: E_1, A and B are both illuminated; E_2, B and C are both illuminated; E_3, all three objects are illuminated. Their probabilities are

$$P(E_1) = (0.62)(0.55) = 0.34100$$

$$P(E_2) = (0.55)(0.71) = 0.39050$$

$$P(E_3) = (0.62)(0.55)(0.71) = 0.24211$$

Events E_1 and E_2 are overlapping, and event E_3 encompasses both E_1 and E_2. By Theorem 5.4, the probability that the requirement is satisfied is

$$0.34100 + 0.39050 - 0.24211 = 0.48939$$

Method 2. We define the following events: E_4, A and B are illuminated but C is not; E_5, B and C are illuminated but A is not. Their probabilities are

$$P(E_4) = (0.62)(0.55)(1 - 0.71) = 0.09889$$

$$P(E_5) = (1 - 0.62)(0.55)(0.71) = 0.14839$$

Events E_3, E_4, and E_5 are mutually exclusive, and the requirement is satisfied if any one of these events occurs. Therefore, by Theorem 5.2, the probability that the requirement is satisfied is

$$0.24211 + 0.09889 + 0.14839 = 0.48939$$

5.4 DEPENDENT TRIALS AND CONDITIONAL PROBABILITY

Consider that two trials, T_1 and T_2, will be performed in the sequence indicated. If the outcome of T_1 influences the outcome of T_2, then T_2 is *dependent* on T_1. For example, assume the following: A case contains three type A units and five type B units; two units will be drawn at random and discarded. What is the probability that the second unit that is drawn is type A? The answer depends on whether the first unit that is drawn is type A or type B. Therefore, the second drawing is dependent on the first.

Let E_1 and E_2 denote events that may result from T_1 and T_2, respectively. If the probability of E_2 is calculated on the assumption that E_1 has occurred, the result is called the *conditional probability* of E_2.

Theorem 5.5. Assume that there will be two trials, the second dependent on the first. The probability that the first trial will yield an event E_1 and the second trial will yield an event E_2 is the product of their respective probabilities, where the probability of E_2 is calculated on the premise that E_1 has occurred.

Let $P(E_2 \mid E_1)$ denote the probability that E_2 will occur, given that E_1 has occurred. Theorem 5.5 may be expressed symbolically in this manner:

$$P(E_1 \text{ and } E_2) = P(E_1) \times P(E_2 \mid E_1)$$

Events E_1 and E_2 constitute a *chain* of events, and Theorem 5.5 can be extended to chains consisting of any number of individual events. The probability of the rth event in the chain is based on the assumption that the preceding $r - 1$ events that were specified have in fact occurred.

Theorem 5.5 also applies to situations where a given event can occur *only* if all the preceding events in the chain have occurred. For example, assume again that a case contains type A and type B units. A unit will be drawn at random and discarded, the process being continued until a type B unit is found. What is the probability that the fifth unit that is drawn is type B? This event can occur only if the first four drawings yield type A units.

EXAMPLE 5.16 An outdoor festival is scheduled to be held on Monday. However, if it rains on Monday, the festival will be held on Tuesday. If it rains on both Monday and Tuesday, the festival will be held on Wednesday. If it rains on all three days, the festival will be canceled. The probability of rain is 0.24 for Monday, 0.35 for Tuesday, and 0.18 for Wednesday. Compute the following: the probabilities concerning the day on which the festival will be held; the probability that the festival will be canceled.

SOLUTION

$$P(\text{Monday}) = 1 - 0.24 = 0.76000$$

The festival will be held on Tuesday if it rains on Monday but not on Tuesday. Then

$$P(\text{Tuesday}) = (0.24)(1 - 0.35) = 0.15600$$

The festival will be held on Wednesday if it rains on Monday and Tuesday but not on Wednesday. Then

$$P(\text{Wednesday}) = (0.24)(0.35)(1 - 0.18) = 0.06888$$

$$P(\text{canceled}) = (0.24)(0.35)(0.18) = 0.01512$$

Since it is a certainty that the festival will be held on one of these three days or it will be canceled, the four probabilities we have calculated must total 1, and they do.

EXAMPLE 5.17 A case is known to contain three type A and eight type B units. A type A unit is required to fill an order. Because the units were not properly labeled, it will be necessary to draw a unit at random, examine it, and discard it, the process being repeated until a type A unit is found. What is the probability that the number of units that are drawn will be four?

SOLUTION The number of units in the case is originally 11, but each drawing reduces the number of units that remain in the case by 1. The specified event occurs if the first three units that are drawn are all type B and the fourth unit is type A.

The probability that the first unit will be type B is 8/11. Assume that such is the case. Ten units remain in the case, and seven of these are type B. Therefore, the probability that the second unit will be type B is 7/10. Continuing in this manner, we find that the probability that a type A unit will be found on the fourth drawing is

$$\frac{8}{11}\,\frac{7}{10}\,\frac{6}{9}\,\frac{3}{8} = \frac{7}{55} = 0.1273$$

EXAMPLE 5.18 Boxes A, B, and C each contain 10 objects. Each object is either red or green, and the color composition in each box is as follows: box A, 4 red and 6 green; box B, 5 red and 5 green; box C, 8 red and 2 green. An object will be drawn at random from box A and placed in box B. Then an object will be drawn at random from box B and placed in box C. Finally, an object will be drawn at random from box C. What is the probability that a red object will be drawn from each box?

SOLUTION The probability that a red object will be drawn from A is 4/10. If this event does occur, box B will then contain 6 red and 5 green objects. Therefore, the probability that a red object will be drawn from B is 6/11. If this event does occur, box C will then contain 9 red and 2 green objects. Therefore, the probability that a red object will be drawn from C is 9/11. The probability that this chain of events will occur is

$$\frac{4}{10}\frac{6}{11}\frac{9}{11} = \frac{216}{1210} = 0.1785$$

EXAMPLE 5.19 With reference to Example 5.18, what is the probability that a green object will be drawn from A, a green object from B, and a red object from C?

SOLUTION Proceeding as before, we obtain this probability:

$$\frac{6}{10}\frac{6}{11}\frac{8}{11} = \frac{288}{1210} = 0.2380$$

EXAMPLE 5.20 After it has been formed, a manufactured article is sent to two departments for chemical treatment. In department 1, it is treated with either chemical A or chemical B; in department 2, it is treated with either chemical C or chemical D. Which chemical it receives in department 2 is determined in part by the specific chemical it received in department 1. The probability that an article will receive chemical A is 0.64, and the probability that it will receive both chemicals A and C is 0.29. What is the probability that an article that received chemical A will also receive chemical C?

SOLUTION Theorem 5.5 yields

$$P(\text{A and C}) = P(\text{A}) \times P(\text{C} \mid \text{A})$$

$$P(\text{C} \mid \text{A}) = \frac{0.29}{0.64} = 0.4531$$

EXAMPLE 5.21 Bodies A, B, C, and D are placed in a row, in that order. If one body is set in motion, it may strike the following body and thereby set the latter in motion. The probability is 0.82 that A will strike B, 0.70 that B will strike C, and 0.65 that C will strike D. Body A is set in motion. Construct every possible chain of events, and compute the probability corresponding to each chain.

SOLUTION The possible chains are as follows:

1. Body A fails to strike B. The probability is

$$1 - 0.82 = 0.1800$$

2. Body A strikes B; B fails to strike C. The probability is

$$(0.82)(1 - 0.70) = 0.2460$$

3. Body A strikes B; B strikes C; C fails to strike D. The probability is

$$(0.82)(0.70)(1 - 0.65) = 0.2009$$

4. The motion of A is transmitted to D. The probability is

$$(0.82)(0.70)(0.65) = 0.3731$$

This set of chains encompasses all possibilities. Therefore, the sum of their probabilities must be 1, and we find that such is the case.

In some instances, the probability corresponding to a given chain of events can be obtained by an alternative procedure, in which we view the events in the chain as a permutation or combination. We shall illustrate this procedure.

EXAMPLE 5.22 Solve Example 5.17 in an alternative manner.

SOLUTION Assume the following: All units of a given type are distinguishable from one another on the basis of some other characteristic; precisely four units are drawn, regardless of their type. Each set of units that is drawn represents a permutation of 11 units taken 4 at a time, and the number of possible permutations is $11 \times 10 \times 9 \times 8 = 7920$. Now consider a permutation in which the first three units are type B and the fourth unit is type A. The number of ways in which this permutation can be formed is $8 \times 7 \times 6 \times 3 = 1008$. Therefore, the probability that the true permutation will have the specified form is $1008/7920 = 0.1273$.

EXAMPLE 5.23 In a game of chance, the player selects four integers from 1 to 9, inclusive, and an official then selects three such integers. If the two sets of integers coincide, the player wins. What is the probability that the player will win? Express the result as a fraction.

SOLUTION

Method 1. The probability that the first integer selected by the official is one that is in the player's set is 4/9. If this integer is in the player's set, the probability that the second integer selected by the official is also in the player's set is 3/8. Continuing in this manner, we find that the probability that the player will win is

$$\frac{4}{9}\,\frac{3}{8}\,\frac{2}{7}\,\frac{1}{6} = \frac{1}{126}$$

Method 2. Since the sequence in which the player selects the integers is immaterial, each set of integers is a combination of 9 integers taken 4 at a time. The number of possible combinations is $C(9,4) = 9!/(4!\,5!) = 126$. Therefore, the probability that the player will win is 1/126.

5.5 RELATIONSHIP BETWEEN PROBABILITY AND RELATIVE FREQUENCY

Assume that n independent trials of identical type were performed and that an event E occurred r times. In accordance with the definition in Art. 4.1, the *relative frequency* of E is r/n.

Now consider that the number of trials n increases beyond bound. It is logical to assume that as it does so the relative frequency of E approaches $P(E)$ as a limit. For example, assume that a given trial may produce an event E and that the probability that E will occur is 0.38. If we repeat this trial endlessly, we assume that E will occur 38 percent of the time. Thus, we may consider that probability equals relative frequency *in the long run.* This relationship affords a relatively simple and practical method of solving numerous problems in probability. We shall illustrate the technique.

EXAMPLE 5.24 A firm manufactures a standard part, and it has three alternative machines, A, B, and C, available for this purpose. The probability that a part selected at random was produced by a particular machine is as follows: A, 45 percent; B, 30 percent; C, 25 percent. The probability that a part produced by a machine is defective is as follows: A, 3.2 percent; B, 4.1 percent; C, 5.9 percent. What is the probability that a part selected at random is defective?

SOLUTION Arbitrarily set the number of parts manufactured in a given period equal to 100,000. Equating probability to relative frequency in the long run, we obtain the following:

$$\begin{aligned}
\text{No. defectives produced by A} &= 100{,}000(0.45)(0.032) = 1440\\
\text{No. defectives produced by B} &= 100{,}000(0.30)(0.041) = 1230\\
\text{No. defectives produced by C} &= 100{,}000(0.25)(0.059) = \underline{1475}\\
\text{Total no. defectives} &= 4145
\end{aligned}$$

Then $$P(\text{defective}) = 4145/100{,}000 = 4.145 \text{ percent}$$

We may interpret this result as the *mean incidence* of defectives.

EXAMPLE 5.25 A firm purchases cables for use in its operations. They are procured from three suppliers, in these proportions: 40 percent from Company A, 35 percent from Company B, and 25 percent from Company C. The probability that a cable procured from a particular supplier is defective is as follows: Company A, 5 percent; Company B, 9 percent; Company C, 10 percent. If a cable is found to be defective, what is the probability that it was supplied by Company A?

SOLUTION Conventionally, a problem of this type is solved by a principle known as *Bayes' theorem.* However, we shall solve it in a simple manner by equating the probability that a cable is defective to the actual incidence of defectives in the long run. Arbitrarily set the number of cables that were purchased to 10,000. Then

$$\begin{aligned}
\text{No. defectives procured from A} &= 10{,}000(0.40)(0.05) = 200\\
\text{No. defectives procured from B} &= 10{,}000(0.35)(0.09) = 315\\
\text{No. defectives procured from C} &= 10{,}000(0.25)(0.10) = \underline{250}\\
\text{Total no. defectives} &= 765
\end{aligned}$$

It follows that the probability that the defective cable came from Company A is $200/765 = 0.2614$.

EXAMPLE 5.26 A box contains spheres and cubes, and each object is yellow or blue. If an object is drawn at random, the probability is 62 percent that it is a sphere and 38 percent that it is a cube. The probability that a sphere is yellow is 71 percent, and the probability that a cube is yellow is 35 percent. An object has been drawn at random, and it is blue. What is the probability that it is a sphere?

SOLUTION Arbitrarily set the total number of objects in the box equal to 10,000. The number of objects that are spherical and blue is

$$10{,}000(0.62)(1-0.71)=1798$$

The number of objects that are cubical and blue is

$$10{,}000(0.38)(1-0.35)=2470$$

The total number of blue objects is $1798+2470=4268$. Therefore, the probability that the object that was drawn is spherical is $1798/4268=0.4213$.

EXAMPLE 5.27 Solve Example 5.20 by equating probability to relative frequency.

SOLUTION Arbitrarily set the number of articles that are manufactured to 100. The number of articles that receive chemical A is $100(0.64)=64$, and the number that receive both A and C is $100(0.29)=29$. Therefore, of the articles that receive A, the proportion that then receive C is $29/64=0.4531$, and this is the required probability.

5.6 DEFINITION OF A PROBABILITY DISTRIBUTION

Let X denote a discrete random variable. All the values that X can assume and their respective probabilities constitute the *probability distribution* of X. Since it is a certainty that X will assume one of its possible values, the sum of these probabilities is 1.

EXAMPLE 5.28 A case contains five type S units and two type T units. A unit will be drawn at random from the case and discarded until a type T unit has been drawn. If X denotes the number of drawings that are required, establish the probability distribution of X.

SOLUTION The procedure is similar to that in Example 5.17, where we assigned a specific value of X (namely, 4). The probability that the first unit is type S is 5/7 and the probability that it is type T is 2/7. If the first unit was type S, the probability that the second unit will also be type S is 4/6 and the probability that it will be type T is 2/6. Since we start with five type S units, the maximum number of possible drawings is 6. Assigning specific values to X, we obtain the following results:

$$P(X=1)=\frac{2}{7}=0.28571 \qquad P(X=2)=\frac{5}{7}\,\frac{2}{6}=0.23810$$

$$P(X=3)=\frac{5}{7}\,\frac{4}{6}\,\frac{2}{5}=0.19048 \qquad P(X=4)=\frac{5}{7}\,\frac{4}{6}\,\frac{3}{5}\,\frac{2}{4}=0.14286$$

$$P(X=5)=\frac{5}{7}\,\frac{4}{6}\,\frac{3}{5}\,\frac{2}{4}\,\frac{2}{3}=0.09524 \qquad P(X=6)=\frac{5}{7}\,\frac{4}{6}\,\frac{3}{5}\,\frac{2}{4}\,\frac{1}{3}\,\frac{2}{2}=0.04762$$

These probabilities total 1, as they must.

5.7 PROPERTIES OF A PROBABILITY DISTRIBUTION

Consider that a trial is performed repeatedly to yield m values of a random variable X. This set of values has an arithmetic mean $\bar{X}$ and a standard deviation s, these terms being defined in Arts. 4.5 and 4.11, respectively. Now consider that the number of values m increases beyond bound. The limiting values approached by $\bar{X}$ and s are termed the arithmetic mean and standard deviation, respectively, of the *probability distribution* of X, and they are denoted by μ and σ, respectively.

Equations (4.1) and (4.7) are adapted to the present case by replacing relative frequency f/n with probability $P(X)$. The results are as follows:

$$\mu = \sum X[P(X)] \tag{5.2}$$

$$\sigma = \sqrt{\sum (X - \mu)^2[P(X)]} \tag{5.3}$$

The arithmetic mean of a probability distribution is also called the *expected value* of the variable, and it is then denoted by $E(X)$.

EXAMPLE 5.29 An individual has purchased a ticket that entitles her to participate in a game of chance. The sums that may be won and their respective probabilities are recorded in Table 5.1. What is the player's expected winning?

TABLE 5.1

Prospective winning, $		Probability
0		0.78
10		0.15
25		0.04
50		0.02
75		0.01
	Total	1.00

SOLUTION

$$\text{Expected winning} = 10(0.15) + 25(0.04) + 50(0.02) + 75(0.01)$$
$$= \$4.25$$

The significance of this result is the following: If the number of times this game is played increases beyond bound, the average amount won in a single play approaches $4.25 as a limit.

EXAMPLE 5.30 With reference to Example 5.28, find the expected number of drawings (to three decimal places).

SOLUTION

$$E(X) = 1(0.28571) + 2(0.23810) + 3(0.19048) + 4(0.14286) + 5(0.09524) + 6(0.04762)$$
$$= 2.667$$

EXAMPLE 5.31 An experiment will be performed repeatedly until a required result is obtained. Let X denote the number of times it will be necessary to perform the experiment. The probability distribution of X is as follows:

$$P(X = 1) = 0.23 \qquad P(X = 2) = 0.28 \qquad P(X = 3) = 0.21$$

$$P(X = 4) = 0.16 \qquad P(X = 5) = 0.09 \qquad P(X = 6) = 0.03$$

The cost of performing an experiment is estimated to be \$10 for the first experiment, \$7 for the second experiment, \$5 for the third experiment, and \$4 for each subsequent experiment. What is the expected cost of the entire set of experiments?

SOLUTION Let C denote this cost. The calculations are performed in Table 5.2. First, we record the value of C corresponding to every possible value of X. For example, if $X = 5$, $C = 10 + 7 + 5 + 2 \times 4 = \$30$. The value of C can range from \$10 (when $X = 1$) to \$34 (when $X = 6$). We then record the probability of every possible value of C (which equals the probability of the corresponding value of X), and sum the products. The result is $E(C) = \$19.56$.

TABLE 5.2

X	C, \$	$P(C)$	$C[P(C)]$, \$
1	10	0.23	2.30
2	$10 + 7 = 17$	0.28	4.76
3	$17 + 5 = 22$	0.21	4.62
4	$22 + 4 = 26$	0.16	4.16
5	$26 + 4 = 30$	0.09	2.70
6	$30 + 4 = 34$	0.03	1.02
		Total	19.56

Probability distributions tend to fall into certain categories, and we shall investigate one category that applies to discrete variables.

5.8 BINOMIAL DISTRIBUTION

The possible outcomes of a trial can be divided into two sets: those that yield a given event, and those that fail to do so. Consider that a given trial is performed n times under identical conditions. The probability that an event E will occur in a given trial is the same for all trials. Let X denote the number of times that E occurs as a result of the n trials. For a reason that will soon become apparent, the probability distribution of X is called *binomial*. We shall first consider a specific variable that has a binomial distribution.

EXAMPLE 5.32 A case contains 5 type A and 9 type B units. A unit will be drawn at random from the case and replaced with one of identical type, the process being repeated until 8 units have been drawn. Let X denote the number of type A units that are drawn. Determine the probability that $X = 3$.

SOLUTION Each drawing has two possible outcomes: A and B. As a result of the replacement, each drawing is independent of the preceding drawings. Therefore, the probability distribution of X is binomial. The probabilities associated with an individual drawing are these: $P(A) = 5/14$; $P(B) = 9/14$.

Consider this acceptable set of outcomes: A-A-A-B-B-B-B-B. The probability of obtaining this set is $(5/14)^3(9/14)^5$. We can rearrange the 3 A's and 5 B's to form another acceptable set of outcomes, and the probability of the new set will be the same. By an extension of Eq. (1.22), the number of possible arrangements is $8!/(3!5!)$. Alternatively, we may consider that the positions occupied by the 3 A's constitute a combination of 8 positions taken 3 at a time. By Eq. (1.24), the number of possible combinations is $C(8,3) = 8!/(3!5!)$. Then

$$P(X = 3) = \frac{8!}{3!5!}\left(\frac{5}{14}\right)^3\left(\frac{9}{14}\right)^5 = 0.2801$$

We shall now generalize on the basis of Example 5.31. Let

P = probability given event will occur on single trial

Q = probability given event will fail to occur on single trial $= 1 - P$

n = number of independent trials

X = number of times given event occurs

Applying the notation for binomial coefficients given in Art. 1.16, we have

$$P(X) = [C(n,X)]P^X Q^{n-X} = \frac{n!}{X!(n-X)!}P^X Q^{n-X} = \binom{n}{X}P^X Q^{n-X} \tag{5.4}$$

5.9 DEFINITIONS AND NOTATION PERTAINING TO MARKOV CHAINS

Assume the following: A given trial will be performed repeatedly, the outcome of each trial is directly influenced by the outcome of the preceding trial, and the strength of this influence remains unchanged as the number of trials increases. A trial having this characteristic is called a *Markov process*, and the set of successive outcomes resulting from the repeated trials is termed a *Markov chain*.

Let E and F denote two possible outcomes of a Markov process. The probability that outcome E will be followed by outcome F is termed a *transition probability*. The symbol E_n denotes that the nth outcome in the chain is E, and $P(E_n)$ denotes the probability that the nth outcome truly will be E. We shall discover that $P(E_n)$ varies with n but approaches a limiting value as n becomes infinite. The value of $P(E_n)$ corresponding to a finite value of n is termed a *transient probability*, and the limiting value approached by $P(E_n)$ is termed a *steady-state probability*.

We shall confine our investigation to Markov processes that have only two possible outcomes, and we shall label them A and B.

5.10 TRANSIENT PROBABILITIES OF A MARKOV CHAIN

Example 5.33 illustrates how the initial transient probabilities are calculated.

EXAMPLE 5.33 A Markov process has the following transition probabilities: The probability that outcome A will be followed by another A is 0.75, and the probability that outcome B will be followed by another B is 0.60. Calculate the probabilities corresponding to the first four outcomes.

SOLUTION Refer to Table 5.3, where the probabilities are arranged in the form of a *transition matrix*.

TABLE 5.3 Transition Matrix

	*n*th outcome	
$(n-1)$th outcome	A_n	B_n
A_{n-1}	0.75	0.25
B_{n-1}	0.40	0.60

From Table 5.3, we obtain the following:

$$P(A_n) = 0.75P(A_{n-1}) + 0.40P(B_{n-1})$$

Replacing $P(B_{n-1})$ with $1 - P(A_{n-1})$ and simplifying, we obtain

$$P(A_n) = 0.35P(A_{n-1}) + 0.40 \qquad (a)$$

Setting $n = 2$ and applying Eq. (*a*), we obtain

$$P(A_2) = 0.35P(A_1) + 0.40 \qquad (b)$$

Now setting $n = 3$, applying Eq. (*a*), and replacing $P(A_2)$ with its expression in Eq. (*b*), we obtain

$$P(A_3) = 0.1225P(A_1) + 0.54 \qquad (c)$$

Continuing this process, we obtain

$$P(A_4) = 0.0429P(A_1) + 0.589 \qquad (d)$$

Assume that the initial outcome is A. Setting $P(A_1) = 1$ and substituting in the foregoing equations, we obtain these results:

$$P(A_2) = 0.7500 \qquad P(A_3) = 0.6625 \qquad P(A_4) = 0.6319$$

Now assume that the initial outcome is B. Setting $P(A_1) = 0$ and substituting in the foregoing equations, we obtain these results:

$$P(A_2) = 0.4000 \qquad P(A_3) = 0.5400 \qquad P(A_4) = 0.5890$$

EXAMPLE 5.34 With reference to Example 5.33, assume that the initial outcome is B and the fourth outcome is A. Compute the probability corresponding to each chain that has these properties.

SOLUTION Consider the chain B-A-A-A. If the initial outcome is B, the probability that the second outcome will be A is 0.40. If the second outcome was in fact A, the probability that the third outcome will be A is 0.75. Continuing in this manner and applying Theorem 5.5, we obtain the following probabilities pertaining to the indicated chains:

$$P(\text{B-A-A-A}) = (0.40)(0.75)(0.75) = 0.225$$

$$P(\text{B-B-A-A}) = (0.60)(0.40)(0.75) = 0.180$$

$$P(\text{B-A-B-A}) = (0.40)(0.25)(0.40) = 0.040$$

$$P(\text{B-B-B-A}) = (0.60)(0.60)(0.40) = 0.144$$

These probabilities total 0.589. Thus, if the initial outcome is B, the probability that the fourth outcome will be A, regardless of what the intervening outcomes may be, is 0.589. This result coincides with that obtained in Example 5.33.

5.11 STEADY-STATE PROBABILITIES OF A MARKOV CHAIN

In Example 5.33, Eqs. (*b*), (*c*), and (*d*) reveal that the influence of the initial outcome becomes progressively more tenuous as the number of trials n becomes larger. Moreover, the calculations that yield these equations reveal that this influence vanishes completely as n increases beyond bound. Thus, as n becomes infinite, $P(\text{A}_n)$ and $P(\text{B}_n)$ approach limiting values that are governed solely by the transition probabilities. As previously stated, these limiting values are termed *steady-state probabilities*.

EXAMPLE 5.35 With reference to Example 5.33, find the limiting values approached by $P(\text{A}_n)$ and $P(\text{B}_n)$ as n becomes infinite.

SOLUTION Let $P^*(\text{A})$ and $P^*(\text{B})$ denote these limits. The simplest method of obtaining these values is the following: In Eq. (*a*), replace both $P(\text{A}_n)$ and $P(\text{A}_{n-1})$ with $P^*(\text{A})$ and solve the resulting equation to obtain the following:

$$P^*(\text{A}) = 0.35P^*(\text{A}) + 0.40 \qquad P^*(\text{A}) = 0.6154$$

We now formulate the following equation, which is analogous to Eq. (*a*):

$$P(\text{B}_n) = 0.35P(\text{B}_{n-1}) + 0.25$$

Proceeding as before, we obtain

$$P^*(\text{B}) = 0.35P^*(\text{B}) + 0.25 \qquad P^*(\text{B}) = 0.3846$$

As a test, we observe that $P^*(\text{A})$ and $P^*(\text{B})$ total 1, as they must.

CHAPTER 6

PLANE GEOMETRY AND TRIGONOMETRY

6.1 PLANE GEOMETRY

While a comprehensive exposition of geometry lies beyond the scope of this book, we shall present the definitions and principles that are most frequently applied in mathematics and engineering.

6.1.1 Angles and Lines

When a line rotates about a fixed point until it returns to its original position, it generates an angle of 360°. An angle of 180° is a *straight angle*, and an angle of 90° is a *right angle*. An angle is *acute* if it lies between 0 and 90°, and it is *obtuse* if it lies between 90° and 180°. Two angles are *supplementary* to each other if their sum is 180°, and they are *complementary* to each other if their sum is 90°.

The symbol $\angle$ denotes *angle*, the symbol $\perp$ means *is perpendicular to*, and the symbol $\|$ means *is parallel to.*

The angles in Fig. 6.1 have these relationships: $a' \perp a$ and $b' \perp b$. Therefore, angles A and B are equal, and angles A and C are supplementary.

If two curves, or a curve and a straight line, touch each other at only one point, the lines are *tangent* to each other.

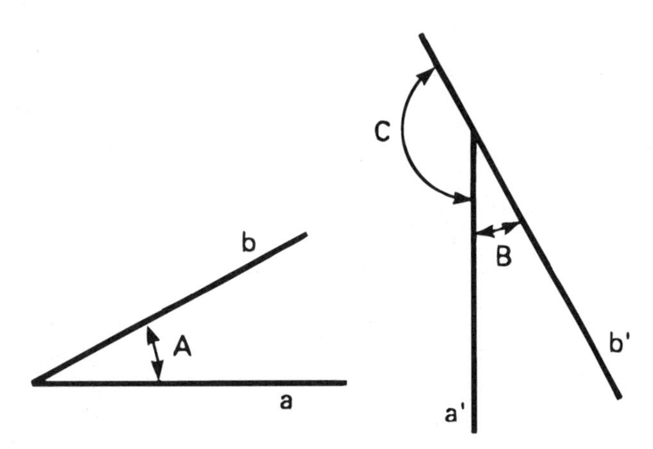

FIGURE 6.1 Angles with mutually perpendicular sides.

6.1.2 Polygons

A *polygon* is a closed figure formed by straight lines. The following are special types of polygons: a triangle, with three sides; a quadrilateral, with four sides; a pentagon, with five sides; an octagon, with eight sides; a decagon, with 10 sides. A *regular* polygon is one in which all sides are equal. Its interior angles are also equal.

The point at which two adjacent sides of a polygon intersect is a *vertex*. A straight line that connects two points on the circumference of a polygon is a *chord*. The sum of the sides of a polygon is its *perimeter*.

Two polygons are *congruent* if their corresponding sides are equal, and they are *similar* if corresponding sides are unequal but have a constant ratio. Similar polygons have the same shape, and all corresponding angles are equal. Conversely, if all corresponding angles in two polygons are equal, the polygons are either congruent or similar.

A polygon is *convex* if it contains no interior angle greater than 180°, and it is *concave* if at least one interior angle exceeds 180°. A convex polygon closes on itself, and any chord that may be drawn lies wholly within the polygon.

The sum of the interior angles of a triangle is 180°. A polygon having n sides can be divided into n triangles by drawing rays from an arbitrary interior point to its vertices. It follows that the sum of the interior angles of a polygon having n sides is $n - 2$ straight angles.

EXAMPLE 6.1 What is each interior angle of a regular polygon having 15 sides?

SOLUTION The sum of the interior angles is $13 \times 180° = 2340°$. Since all interior angles are equal, each angle is $2340/15 = 156°$.

6.1.3 Quadrilaterals

The interior angle of a quadrilateral is assigned the same designation as the vertex at which it is located. For example, in Fig. 6.2*a*, $\angle BAD$ is called simply $\angle A$.

A *trapezoid* is a quadrilateral in which two sides are parallel to each other, and these sides are called the *bases*. For example, in Fig. 6.2*a*, $AB \parallel DC$. Therefore, this quadrilateral is a trapezoid, and AB and DC are its bases. An *isosceles* trapezoid is one in which the nonparallel sides are equal to each other. For example, with reference to the trapezoid in Fig. 6.2*b*, $AD = BC$; therefore, this trapezoid is isosceles. Angles A and B are equal, and angles C and D are equal.

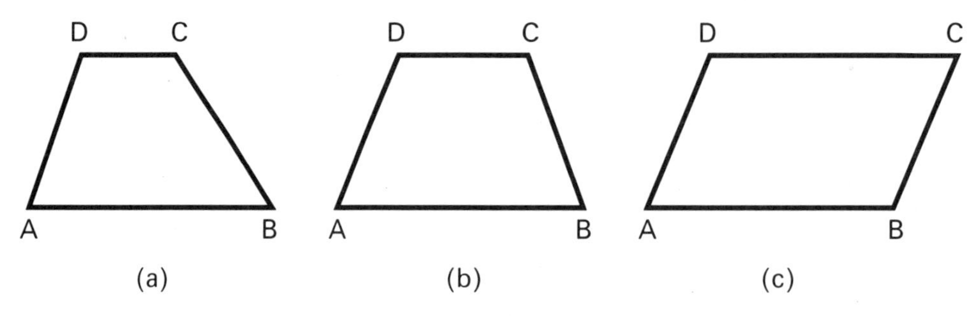

FIGURE 6.2 Special quadrilaterals. (*a*) Trapezoid; (*b*) isosceles trapezoid; (*c*) parallelogram.

A *parallelogram* is a quadrilateral in which two pairs of sides are parallel, as illustrated in Fig. 6.2*c*. The following equalities exist: $AB = DC$; $AD = BC$; $\angle A = \angle C$; $\angle B = \angle D$. A *rectangle* is a parallelogram in which all angles are 90°, and a *rhombus* is a parallelogram in which all sides are equal.

6.1.4 Triangles

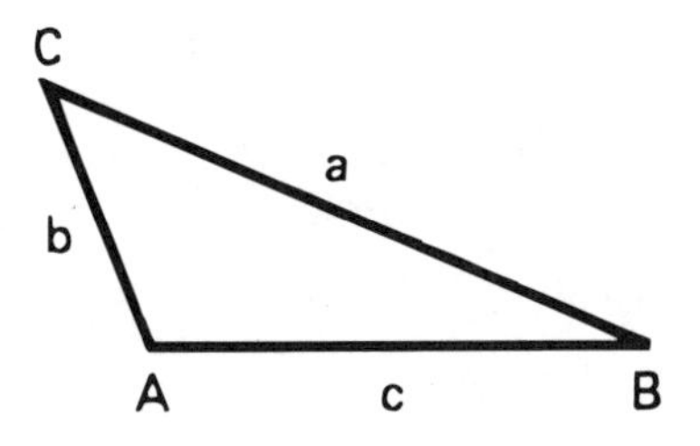

FIGURE 6.3 Method of labeling sides of a triangle.

The side of a triangle is assigned the lowercase letter that corresponds to the opposite vertex, as illustrated in Fig. 6.3. As previously stated, the angles of a triangle have this relationship:

$$A + B + C = 180° \tag{6.1}$$

An *isosceles triangle* is one in which two and only two sides are equal. In Fig. 6.4*a*, $AC = BC$; therefore, this triangle is isosceles. Angles A and B are equal. An *equilateral triangle* is one in which all three sides are equal. All angles are equal, and each angle is 60°. An *oblique* or *scalene triangle* is one in which all three sides are unequal.

A *right triangle* is one that has an angle of 90°, as shown in Fig. 6.4*b*. Side AB is the *hypotenuse*, and sides AC and BC are the *legs*. By the Pythagorean theorem,

$$a^2 + b^2 = c^2 \tag{6.2}$$

In Fig. 6.4*c*, line AB has been prolonged to D. Angle DBC is called an *exterior* angle of the triangle. Since this angle is supplementary to B, it follows from Eq. (6.1) that $\angle DBC = \angle A + \angle C$. We thus arrive at this general principle: An exterior angle of a triangle equals the sum of the two opposite interior angles.

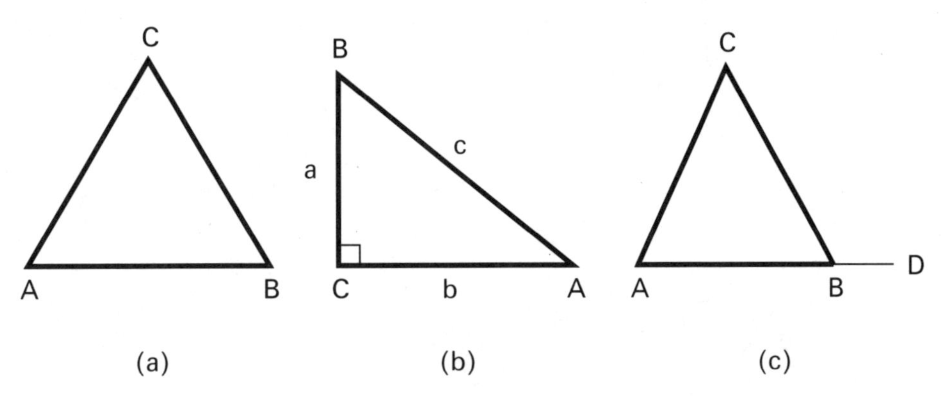

FIGURE 6.4 (*a*) Isosceles triangle; (*b*) right triangle; (*c*) exterior angle of a triangle.

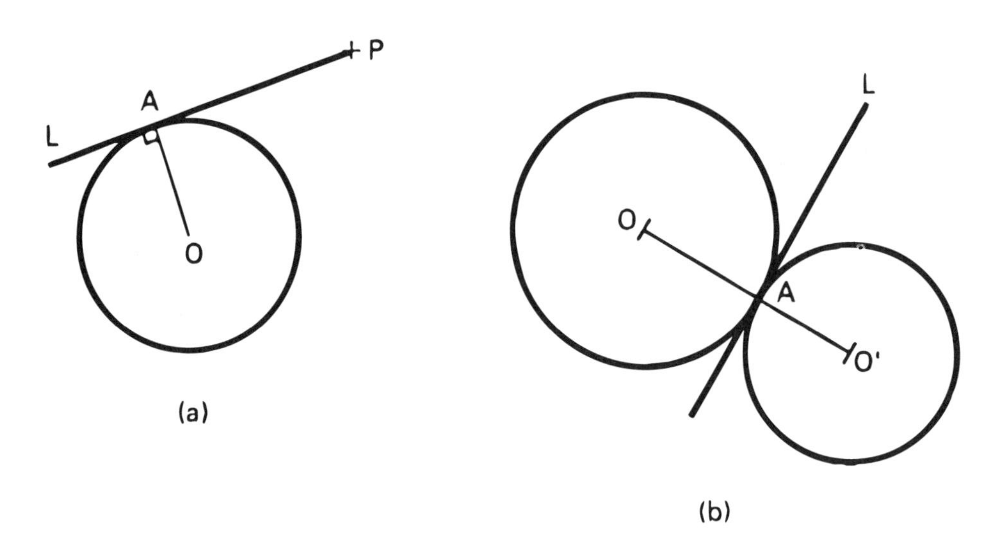

FIGURE 6.5 (*a*) Tangent to a circle; (*b*) tangent circles.

A line drawn from a vertex perpendicular to the opposite side (prolonged if necessary) is an *altitude*, and a line drawn from a vertex to the midpoint of the opposite side is a *median*.

6.1.5 Circles

In Fig. 6.5*a*, O is the center of the circle, P is a point that lies outside the circle, and L is a line through P that is tangent to the circle at A. The radius OA is perpendicular to L. In Fig. 6.5*b*, the circles having their centers at O and O' are tangent to each other at A. The *line of centers* OO' contains A. The line L that is perpendicular to OO' at A is tangent to both circles.

With reference to Fig. 6.6*a*, the arc AB (denoted by $\widehat{AB}$) is assigned the same measure as the angle AOB subtended at the center. Thus, if $\angle AOB = 58°$, $\widehat{AB} = 58°$.

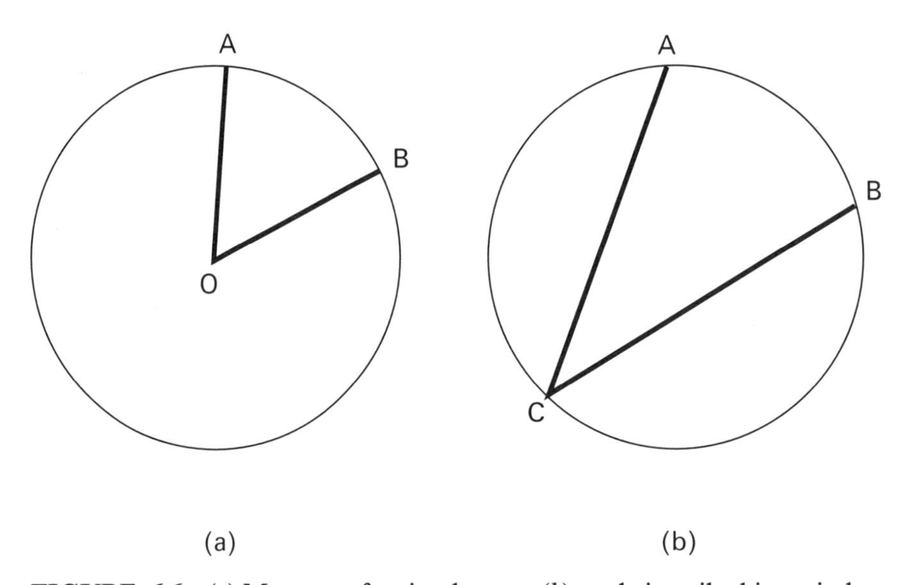

FIGURE 6.6 (*a*) Measure of a circular arc; (*b*) angle inscribed in a circle.

An angle inscribed in a circle is measured by one-half its intercepted arc. For example, in Fig. 6.6*b*, if $\widehat{AB} = 76°$, $\angle ACB = 38°$. It follows that an angle inscribed in a semicircle is a right angle.

The ratio of the circumference of a circle to the diameter is constant, and this constant is denoted by π. Then

$$C = \pi d = 2\pi r \tag{6.3}$$

where C is the circumference, d the diameter, and r the radius of the circle.

6.2 PLANE TRIGONOMETRY

6.2.1 Characteristics of Angles

With reference to Fig. 6.7*a*, consider that line OA on the positive side of the x axis revolves about the origin O. The angle thus generated is positive or negative according to whether the direction of rotation is counterclockwise or clockwise, respectively. Thus, $\angle AOB$ is positive and $\angle AOC$ is negative. In $\angle AOB$, OA and OB are the initial and terminal sides, respectively.

Three units of angular measure are used. One is the degree (°), and another is the grad. A complete revolution generates 360° and 400 grad. The third unit is the radian (rad). With reference to $\angle AOB$ in Fig. 6.7*b*, draw a circular arc having O as center and an arbitrary radius OC. The number of radians in the angle is the ratio of the length of arc CD to the length of OC. Thus, the radian is a dimensionless quantity. By Eq. (6.3), the angle generated in a complete revolution has 2π rad.

The x and y axes divide the plane into quadrants, and these are numbered in the manner shown in Fig. 6.8*a*. If the initial side of an angle lies on the positive side of the x axis, the angle is said to lie in the quadrant in which its terminal side is located. Thus, in Fig. 6.8*b*, $\angle AOB$ lies in the first quadrant and $\angle AOC$ lies in the third quadrant.

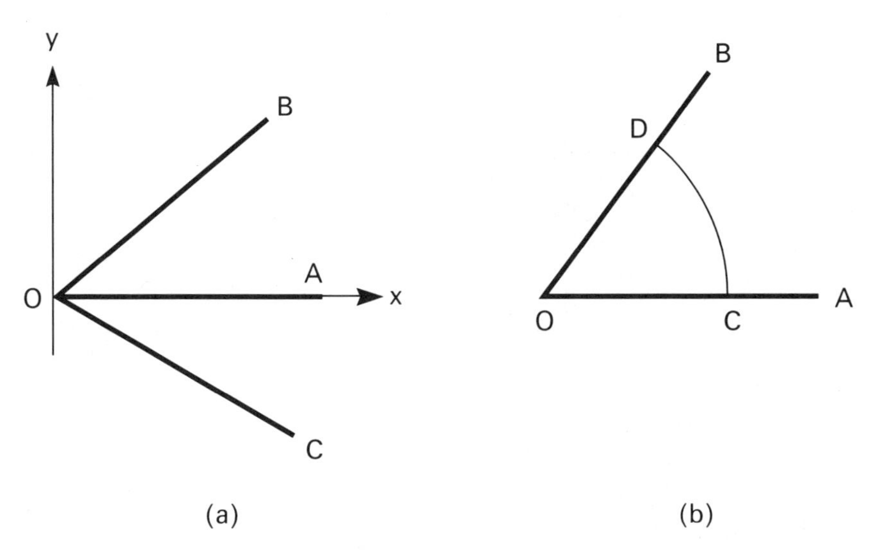

FIGURE 6.7 (*a*) Positive and negative angles; (*b*) radian measure of an angle.

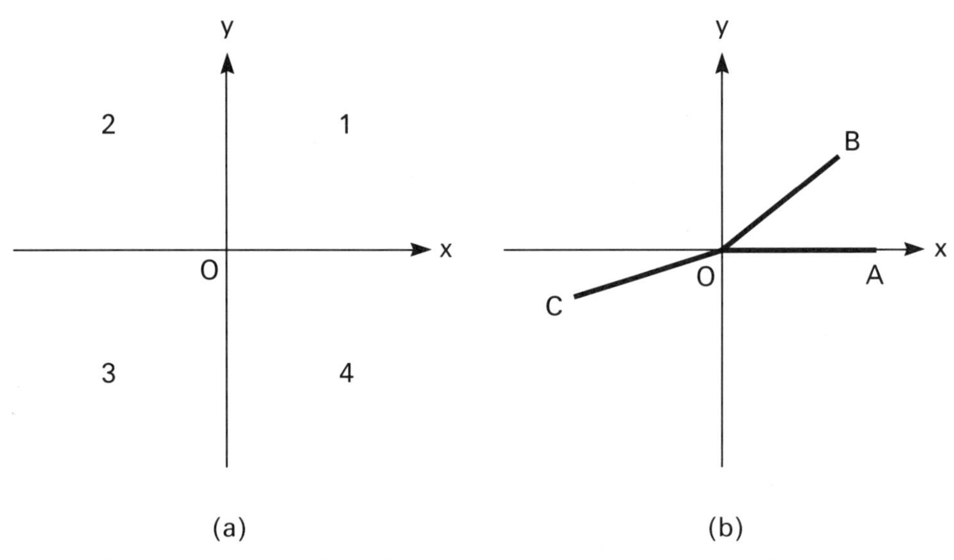

FIGURE 6.8 (*a*) Numbering of quadrants; (*b*) location of an angle by quadrants.

6.2.2 General Expression for an Angle

The appearance of an angle does not divulge how many complete revolutions (if any) occurred during its formation, or the direction in which the generating line moved. However, this information is often significant. Where such is the case, it is necessary to express the size of the angle in *general form* by adding to the apparent size k complete revolutions, where k is an arbitrary integer, positive, negative, or 0. Thus, if the apparent size is $d°$, the size is expressed as $(360k + d)°$; if the apparent size is r rad, the size is expressed as $(2\pi k + r)$ rad. For example, if the apparent size is 65°, the angle may be considered to be 65°, 425°, 785°, −295°, etc. All these angles are *coterminal.*

6.2.3 Functions of an Angle

In Fig. 6.9, let θ denote an angle that has its initial side OA on the positive side of the x axis and its terminal side OB in any quadrant whatever. From an arbitrary point P on OB, drop a perpendicular QP to the x axis. Let $OQ = x$, $QP = y$, and $OP = r$. The functions of the angle are as follows:

$$\sin\theta = \frac{y}{r} \qquad \cos\theta = \frac{x}{r} \qquad \tan\theta = \frac{y}{x}$$

Corresponding to these basic functions are the following reciprocal functions:

$$\csc\theta = \frac{1}{\sin\theta} = \frac{r}{y} \qquad \sec\theta = \frac{1}{\cos\theta} = \frac{r}{x} \qquad \cot\theta = \frac{1}{\tan\theta} = \frac{x}{y}$$

The distance r is considered to be positive, and the algebraic sign of a function of θ depends on the quadrant in which θ is located. Table 6.1 exhibits the algebraic signs of the functions.

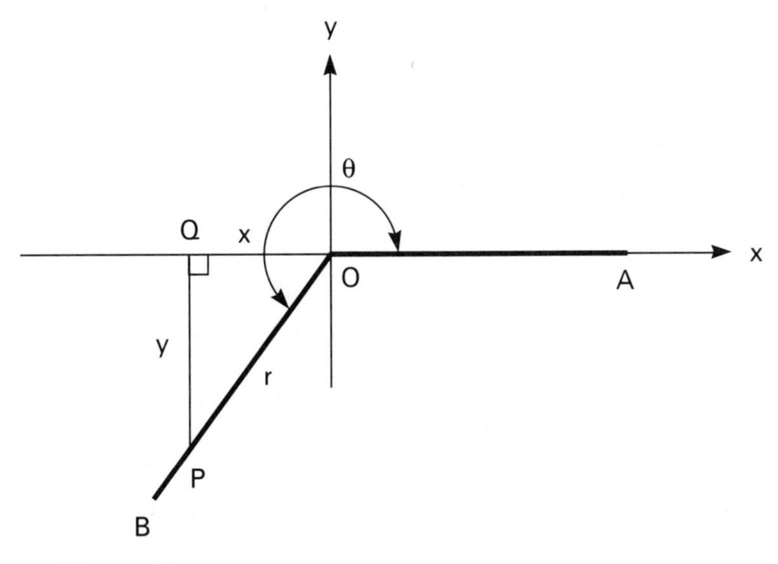

FIGURE 6.9

Assume that a function of an angle is to be raised to the mth power, where $m \neq -1$. The exponent is placed directly after the word. For example $\sin^{3.7}\theta$ means $(\sin\theta)^{3.7}$. However, if $m = -1$, parentheses are mandatory, as in $(\sin\theta)^{-1}$.

TABLE 6.1 Algebraic Signs of Trigonometric Functions

	Quadrant			
Function	1	2	3	4
$\sin\theta$	+	+	−	−
$\cos\theta$	+	−	−	+
$\tan\theta$	+	−	+	−

6.2.4 Basic Relationships among Functions of an Angle

The functions of an angle have the following interrelationships:

$$\tan\theta = \frac{\sin\theta}{\cos\theta} = \frac{\sec\theta}{\csc\theta} \tag{6.4}$$

$$\sin^2\theta + \cos^2\theta = 1 \tag{6.5}$$

$$1 + \tan^2\theta = \sec^2\theta \tag{6.6}$$

$$1 + \cot^2\theta = \csc^2\theta \tag{6.7}$$

If θ is extremely small, $\sin\theta$ and $\tan\theta$ are very close to θ when θ is expressed in radians. Therefore, in practice these functions can often be replaced with θ, or vice versa, without introducing any appreciable error.

6.2.5 Functions of Sum and Difference of Two Angles

Let α and β denote two angles. The functions of their sum and difference are as follows:

$$\sin(\alpha+\beta) = \sin\alpha\cos\beta + \cos\alpha\sin\beta \tag{6.8}$$

$$\cos(\alpha+\beta) = \cos\alpha\cos\beta - \sin\alpha\sin\beta \tag{6.9}$$

$$\tan(\alpha+\beta) = \frac{\tan\alpha + \tan\beta}{1 - \tan\alpha\tan\beta} \tag{6.10}$$

$$\sin(\alpha-\beta) = \sin\alpha\cos\beta - \cos\alpha\sin\beta \tag{6.11}$$

$$\cos(\alpha-\beta) = \cos\alpha\cos\beta + \sin\alpha\sin\beta \tag{6.12}$$

$$\tan(\alpha-\beta) = \frac{\tan\alpha - \tan\beta}{1 + \tan\alpha\tan\beta} \tag{6.13}$$

6.2.6 Functions of Multiples and Submultiples of an Angle

The equations of Art. 6.2.5 yield the following:

$$\sin 2\alpha = 2\sin\alpha\cos\alpha \tag{6.14}$$

$$\cos 2\alpha = \cos^2\alpha - \sin^2\alpha = 1 - 2\sin^2\alpha = 2\cos^2\alpha - 1 \tag{6.15}$$

$$\tan 2\alpha = \frac{2\tan\alpha}{1 - \tan^2\alpha} \tag{6.16}$$

The foregoing equations in turn yield the following:

$$\sin\frac{\alpha}{2} = \pm\sqrt{\frac{1-\cos\alpha}{2}} \tag{6.17}$$

$$\cos\frac{\alpha}{2} = \pm\sqrt{\frac{1+\cos\alpha}{2}} \tag{6.18}$$

$$\tan\frac{\alpha}{2} = \pm\sqrt{\frac{1-\cos\alpha}{1+\cos\alpha}} = \frac{1-\cos\alpha}{\sin\alpha} = \frac{\sin\alpha}{1+\cos\alpha} \tag{6.19}$$

6.2.7 Transformation of Products of Functions

The equations of Art. 6.2.5 also yield the following:

$$\sin\alpha\sin\beta = -\frac{1}{2}\cos(\alpha+\beta) + \frac{1}{2}\cos(\alpha-\beta) \tag{6.20}$$

$$\sin\alpha\cos\beta = \frac{1}{2}\sin(\alpha+\beta) + \frac{1}{2}\sin(\alpha-\beta) \tag{6.21}$$

$$\cos\alpha\cos\beta = \frac{1}{2}\cos(\alpha+\beta) + \frac{1}{2}\cos(\alpha-\beta) \tag{6.22}$$

6.2.8 Inverse Functions

The notation $\theta = \arcsin a$ means that θ is an angle whose sine is a. For example, since $\sin 30° = 0.5$, then $30° = \arcsin 0.5$. The expression $\arcsin a$ has the alternative forms invsin a and $\sin^{-1} a$. Analogous statements apply with reference to the other functions.

Let θ and a denote an angle and a real number, respectively, that are related by a trigonometric function. There is a unique value of a corresponding to a given value of θ. On the other hand, there is an infinite number of values of θ corresponding to a given value of a, and the calculator provides only one value in this set. We shall denote this value by θ_c. Other values of θ may be obtained by applying the following relationships:

$$\sin (180° - \theta) = \sin \theta \qquad \cos (360° - \theta) = \cos \theta$$

$$\tan (180° + \theta) = \tan \theta$$

Moreover, a function of $360° + \theta$ equals the corresponding function of θ.

EXAMPLE 6.2 Applying the restriction $0 < \theta < 360°$, solve the following equations: (*a*) $\theta = \arcsin 0.7$; (*b*) $\theta = \arcsin (-0.7)$; (*c*) $\theta = \arccos 0.4$; (*d*) $\theta = \arccos (-0.4)$; (*e*) $\theta = \arctan 2.1$; (*f*) $\theta = \arctan (-2.1)$; (*g*) $\theta = \text{arcsec}\ 3.2$.

SOLUTION As Table 6.1 discloses, there are two values of θ that satisfy each equation. We shall label these θ_1 and θ_2. With reference to Eq. (*g*) in this example, we have $\cos \theta = 1/(\sec \theta) = 1/3.2 = 0.3125$. The solutions are presented in Table 6.2.

TABLE 6.2 Solutions to Example 6.2

Equation	θ_c, °	θ_1, °	θ_2, °
a	44.43	44.43	$180 - \theta_c = 135.57$
b	−44.43	$180 - \theta_c = 224.43$	$360 + \theta_c = 315.57$
c	66.42	66.42	$360 - \theta_c = 293.58$
d	113.58	113.58	$360 - \theta_c = 246.42$
e	64.54	64.54	$180 + \theta_c = 244.54$
f	−64.54	$180 + \theta_c = 115.46$	$360 + \theta_c = 295.46$
g	71.79	71.79	$360 - \theta_c = 288.21$

6.2.9 Solution of Right Triangle

A triangle consists of six elements: three sides and three angles. If we are given three elements, at least one of which is a side, we can evaluate the three remaining elements, and this process is called *solving the triangle*.

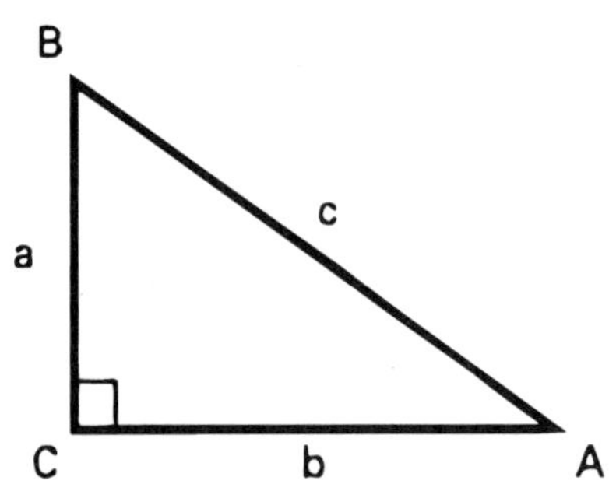

FIGURE 6.10 Right triangle.

Refer to the right triangle in Fig. 6.10. The following relationships exist:

$$\sin A = \cos B = \frac{a}{c} \tag{6.23}$$

$$\cos A = \sin B = \frac{b}{c} \tag{6.24}$$

$$\tan A = \cot B = \frac{a}{b} \tag{6.25}$$

EXAMPLE 6.3 Solve the triangle in which $a = 5$, $A = 36°$, and $C = 90°$.

SOLUTION

$$B = 90° - A = 90° - 36° = 54°$$

Equations (6.23) and (6.25) yield the following:

$$c = \frac{a}{\sin A} = \frac{5}{\sin 36°} = 8.507$$

$$b = \frac{a}{\tan A} = \frac{5}{\tan 36°} = 6.882$$

Equation (6.24) provides a test of these results.

$$\sin B = \frac{b}{c} = \frac{6.882}{8.507} = 0.8090 \qquad \sin 54° = 0.8090$$

Thus, Eq. (6.24) is satisfied, and we find that Eq. (6.2) is also satisfied.

EXAMPLE 6.4 Solve the triangle in which $b = 12.5$, $A = 41°$, and $C = 90°$.

SOLUTION

$$B = 90° - A = 90° - 41° = 49°$$

$$c = \frac{b}{\cos A} = \frac{12.5}{\cos 41°} = 16.563$$

$$a = b \tan A = 12.5 \tan 41° = 10.866$$

Equation (6.23) provides a test of these results.

$$\cos B = \frac{a}{c} = \frac{10.866}{16.563} = 0.6560 \qquad \cos 49° = 0.6561$$

Thus, Eq. (6.23) is satisfied, and we find that Eq. (6.2) is also satisfied.

EXAMPLE 6.5 Solve the triangle in which $a = 10.3$, $c = 16.8$, and $C = 90°$.

SOLUTION

$$\sin A = \cos B = \frac{a}{c} = \frac{10.3}{16.8} = 0.6131$$

$$A = 37.81° \qquad B = 52.19°$$

$$b^2 = c^2 - a^2 = 16.8^2 - 10.3^2 \qquad b = 13.272$$

Equation (6.24) provides a test of these results.

$$\cos A = \sin B = \frac{b}{c} = \frac{13.272}{16.8} = 0.7900$$

$$\cos 37.81^\circ = \sin 52.19^\circ = 0.7900$$

EXAMPLE 6.6 Solve the triangle in which $a = 8.2$, $b = 11.7$, and $C = 90^\circ$.

SOLUTION

$$c^2 = a^2 + b^2 = 8.2^2 + 11.7^2 \qquad c = 14.287$$

$$\tan A = \frac{a}{b} = \frac{8.2}{11.7} = 0.7009$$

$$A = 35.03^\circ \qquad B = 54.97^\circ$$

Equation (6.23) provides a test.

$$\sin A = \cos B = \frac{a}{c} = \frac{8.2}{14.287} = 0.5739$$

$$\sin 35.03^\circ = \cos 54.97^\circ = 0.5740$$

6.2.10 Solution of Oblique Triangle

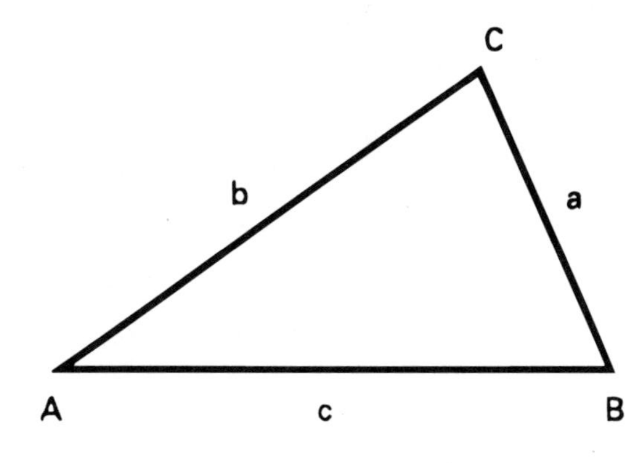

FIGURE 6.11 Oblique triangle.

The oblique triangle in Fig. 6.11 has the following relationships:

$$\frac{a}{\sin A} = \frac{b}{\sin B} = \frac{c}{\sin C} \tag{6.26}$$

$$a^2 = b^2 + c^2 - 2bc \cos A \tag{6.27a}$$

$$b^2 = a^2 + c^2 - 2ac \cos B \tag{6.27b}$$

$$c^2 = a^2 + b^2 - 2ab \cos C \tag{6.27c}$$

Equation (6.26) is known as the *law of sines*, and Eqs. (6.27) constitute the *law of cosines*.

In general, an angle in an oblique triangle can be acute or obtuse. Therefore, when an angle is evaluated by the law of sines, two prospective values of the angle emerge, in accordance with the discussion in Art. 6.2.8. It is then necessary to identify the true value where only one value is possible.

Problems that require the solution of an oblique triangle can be classified on the basis of the given information, in the manner shown in Table 6.3. In solving a triangle, it is helpful to construct the triangle with ruler and compasses by applying the given information. This construction instantly clarifies the problem.

EXAMPLE 6.7 Solve the triangle in which $A = 42.9^\circ$, $B = 31.6^\circ$, and $a = 20.5$.

TABLE 6.3 Classification of Problems on Oblique Plane Triangles

Type	Given information
1	Two angles and one side
2	Two sides and the angle opposite one of them
3	Two sides and the included angle
4	The three sides

SOLUTION This is a type 1 problem.

$$C = 180° - (A + B) = 105.5°$$

$$b = \frac{a \sin B}{\sin A} = 15.780 \qquad c = \frac{a \sin C}{\sin A} = 29.020$$

These results satisfy Eq. (6.27*a*).

EXAMPLE 6.8 Solve the triangle in which $b = 13.4$, $c = 8.6$, and $C = 31.5°$.

SOLUTION This is a type 2 problem. Figure 6.12 reveals that this problem is ambiguous because the construction yields two triangles rather than one. We shall apply the subscripts 1 and 2 to distinguish their values.

$$\sin B = \frac{b \sin C}{c} = 0.8141$$

$$B_1 = 54.5° \qquad B_2 = 125.5°$$

$$A_1 = 180° - (B_1 + C) = 94° \qquad A_2 = 180° - (B_2 + C) = 23°$$

$$a_1 = \frac{c \sin A_1}{\sin C} = 16.419 \qquad a_2 = \frac{c \sin A_2}{\sin C} = 6.431$$

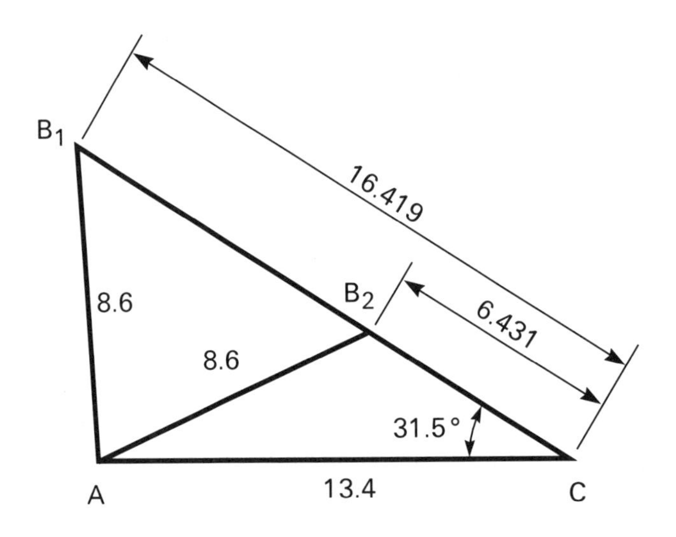

FIGURE 6.12

These results satisfy Eq. (6.27*c*).

EXAMPLE 6.9 Solve the triangle in which $a = 18.3$, $b = 13.9$, and $C = 29°$.

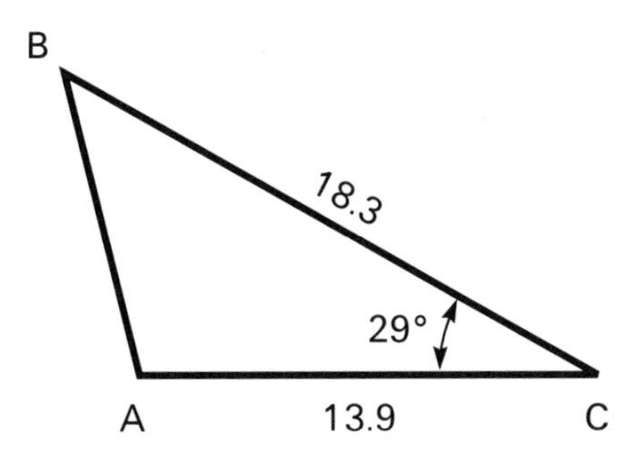

FIGURE 6.13

SOLUTION This is a type 3 problem. Refer to Fig. 6.13. By Eq. (6.27*c*), we obtain $c = 9.118$. Then

$$\sin A = \frac{a \sin C}{c} = 0.9730$$

$$\sin B = \frac{b \sin C}{c} = 0.7391$$

The prospective values of these angles are

$$A_1 = 76.7° \qquad A_2 = 103.3° \qquad B_1 = 47.7° \qquad B_2 = 132.3°$$

Figure 6.13 reveals that the acceptable values are $A = 103.3°$ and $B = 47.7°$. These values satisfy Eq. (6.1).

EXAMPLE 6.10 Solve the triangle in which $a = 9.0$, $b = 15.8$, and $c = 22.1$.

SOLUTION This is a type 4 problem. By the law of cosines,

$$\cos A = \frac{b^2 + c^2 - a^2}{2bc} = 0.9408 \qquad A = 19.8°$$

$$\cos B = \frac{a^2 + c^2 - b^2}{2ac} = 0.8038 \qquad B = 36.5°$$

$$\cos C = \frac{a^2 + b^2 - c^2}{2ab} = -0.5547 \qquad C = 123.7°$$

These results satisfy Eqs. (6.1) and (6.26).

6.2.11 Geometrical Representation of Complex Numbers

In Art. 1.1, we defined imaginary and complex numbers; in Art. 1.9, we stated that $i = \sqrt{-1}$ is the unit of imaginary numbers. We shall now demonstrate that complex numbers can be represented by points on a plane, and we shall then demonstrate that this geometrical representation can be applied advantageously to simplify calculations. This simplified method of calculation is applied extensively in electrical engineering in analyzing circuits with alternating currents.

Let a denoe a real number. In Fig. 6.14, we draw the horizontal and vertical axes shown and then draw a circle having the origin O as center and a as radius. We may say that points A and B on the horizontal axis represent the numbers a and $-a$, respectively. Since $a(-1) = -a$, we may view the process of multiplying a by -1 geometrically as one that causes the radius OA to rotate about O through an angle of 180° in the counterclockwise direction. Since multiplying a by -1 is equivalent to

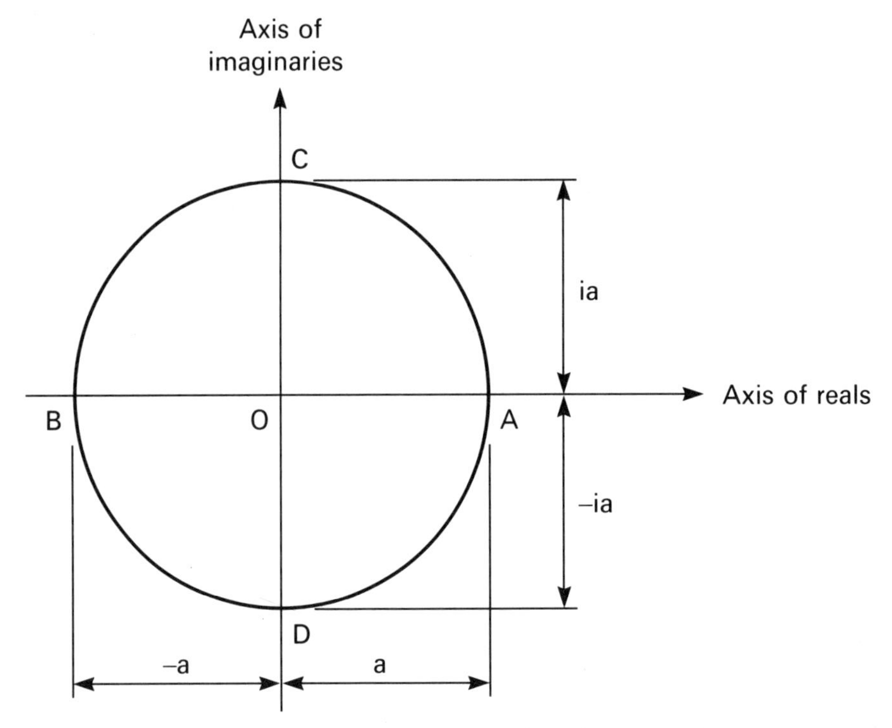

FIGURE 6.14 Representation of real and imaginary numbers.

multiplying a by i twice, we may consider that multiplying a by i *once* causes OA to rotate through an angle of 90° to OC, and multiplying a by i *three times* causes OA to rotate through an angle of 270° to OD. Thus, points C and D on the vertical axis represent the numbers ia and $-ia$, respectively, and it follows that every imaginary number is represented by a point on the vertical axis. Therefore, the horizontal and vertical axes are referred to as the *axis of reals* and the *axis of imaginaries*, respectively.

Going one step further, it is logical to consider that every point in the plane that does not lie on the horizontal or vertical axis represents a *complex* number. For example, consider point P in Fig. 6.15, where $OA = a$ and $OB = ib$. This point rep-

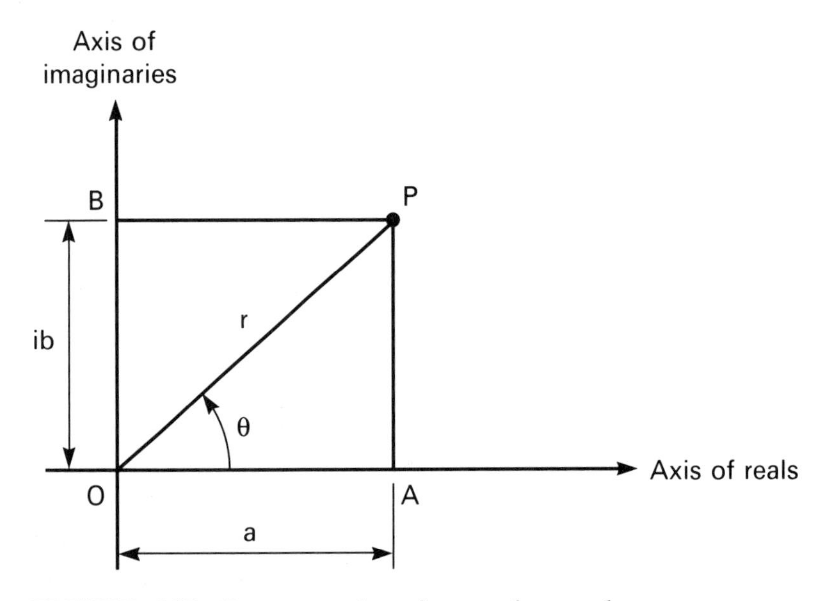

FIGURE 6.15 Representation of a complex number.

resents the complex number $z = a + ib$. The quantities a and b are the *rectangular coordinates* of P.

Now let $OP = r$ and $\angle AOP = \theta$. The quantities r and θ are the *polar coordinates* of P. Then $a = r \cos \theta$ and $b = r \sin \theta$. Therefore, the complex number z represented by point P can be expressed in the following alternative forms:

$$z = a + ib \qquad z = r(\cos \theta + i \sin \theta)$$

The first expression is the *rectangular form* of z, and the second expression is the *polar form*. The polar form is often presented in this shorthand notation:

$$z = r\underline{/\theta}$$

6.2.12 Multiplication and Division of Complex Numbers in Polar Form

We shall operate on the following complex numbers:

$$z_1 = r_1(\cos \theta_1 + i \sin \theta_1) \qquad z_2 = r_2(\cos \theta_2 + i \sin \theta_2)$$

When we multiply these numbers and apply Eqs. (6.8) and (6.9), we obtain this result:

$$z_1 z_2 = r_1 r_2[\cos (\theta_1 + \theta_2) + i \sin (\theta_1 + \theta_2)] \tag{6.28}$$

Similarly, when we divide z_1 by z_2 in the manner described in Art. 1.9, we obtain this result:

$$\frac{z_1}{z_2} = \frac{r_1}{r_2} [\cos (\theta_1 - \theta_2) + i \sin (\theta_1 - \theta_2)] \tag{6.29}$$

In shorthand notation, Eqs. (6.28) and (6.29) can be expressed in this manner:

$$z_1 z_2 = r_1 r_2 \underline{/\theta_1 + \theta_2} \qquad \frac{z_1}{z_2} = \frac{r_1}{r_2} \underline{/\theta_1 - \theta_2}$$

The operation of adding or subtracting complex numbers can be performed readily when the numbers are in rectangular form, but the operation of multiplying or dividing complex numbers can be performed far more efficiently when the numbers are in polar form.

EXAMPLE 6.11 The following complex numbers are given:

$$z_1 = 12 - i5 \qquad z_2 = 8 + i6$$

$$z_3 = -9 + i4 \qquad z_4 = 7 + i12$$

Find $z_1 z_2 z_3/z_4$, and express the result in rectangular form.

SOLUTION Let z_5 denote the result. The calculator enables us to convert the rectangular coordinates of a point to its polar coordinates, and vice versa. However, in the present instance, we shall perform the conversions by applying the relationships among the coordinates, and we recommend that the reader also obtain them by calculator for practice. It is mandatory that each angle be placed in its proper quadrant, and this can be done easily by locating each point in a free-hand sketch.

$$r_1 = \sqrt{12^2 + (-5)^2} = 13 \qquad \theta_1 = \arctan\,[(-5)/12] = -22.620^\circ$$

$$r_2 = \sqrt{8^2 + 6^2} = 10 \qquad \theta_2 = \arctan\,(6/8) = 36.870^\circ$$

$$r_3 = \sqrt{(-9)^2 + 4^2} = 9.849 \qquad \theta_3 = \arctan\,[4/(-9)] = 156.038^\circ$$

$$r_4 = \sqrt{7^2 + 12^2} = 13.892 \qquad \theta_4 = \arctan\,(12/7) = 59.744^\circ$$

Applying Eqs. (6.28) and (6.29), we obtain

$$r_5 = \frac{13 \times 10 \times 9.849}{13.892} = 92.166$$

$$\theta_5 = -22.620^\circ + 36.870^\circ + 156.038^\circ - 59.744^\circ = 110.544^\circ$$

Thus, in polar form,

$$z_5 = 92.166\underline{/110.544^\circ}$$

In rectangular form,

$$z_5 = 92.166(\cos 110.544^\circ + i \sin 110.544^\circ) = -32.34 + i86.31$$

The conversion from polar to rectangular form can of course be obtained by calculator.

6.2.13 DeMoivre's Theorem

Consider that the complex number z is multiplied by itself repeatedly to produce z^n. Equation (6.28) yields the following:

$$z^n = [r(\cos\theta + i \sin\theta)]^n = r^n(\cos n\theta + i \sin n\theta) \tag{6.30}$$

In shorthand notation,

$$(r\underline{/\theta})^n = r^n\underline{/n\theta}$$

This relationship can be extended to include all rational values of n. Equation (6.30) is known as *DeMoivre's theorem*, and it has numerous applications, as we shall now demonstrate.

EXAMPLE 6.12 If $z = 2 - i7$, what is z^6?

SOLUTION The answer can be obtained by repeated multiplication, but Eq. (6.30) provides a more rapid solution. With reference to the point representing z, we have the following:

$$r^2 = 2^2 + (-7)^2 = 53 \qquad r^6 = 53^3 = 148{,}877$$

$$\theta = -74.0546^\circ \qquad 6\theta = -444.3276^\circ$$

Thus, in polar form,

$$z^6 = 148{,}877\underline{/-444.3276^\circ}$$

The corresponding rectangular coordinates are 14,715 and −148,148. Therefore, in rectangular form,

$$z^6 = 14{,}715 - i148{,}148$$

EXAMPLE 6.13 Find the cube roots of $-4 + i3$.

SOLUTION A number has n nth roots; therefore, a number has three cube roots. The polar coordinates of the point representing this number are 5 and 143.13°. However, to obtain all three cube roots, we must express the angle in the general form described in Art. 6.2.2. Let r and θ denote the polar coordinates of the point representing a cube root. Equation (6.30) yields

$$r^3\underline{/3\theta} = 5\underline{/360k^\circ + 143.13^\circ}$$

Then

$$r = 5^{1/3} = 1.7100$$

Now setting k equal to 0, 1, and 2 in turn and then dividing by 3, we obtain the values of θ recorded in Table 6.4. Converting polar coordinates to rectangular coordinates, we obtain the values of a and b shown in the table. In rectagular form, the three cube roots of $-4 + i3$ are as follows:

$$z_1 = 1.1506 + i1.2650 \qquad z_2 = -1.6708 + i0.3640$$

$$z_3 = 0.5202 - i1.6290$$

The results can be verified by raising each of these numbers to the third power.

TABLE 6.4

k	θ, °	a	b
0	$\frac{143.13}{3} = 47.71$	1.1506	1.2650
1	$\frac{503.13}{3} = 167.71$	−1.6708	0.3640
2	$\frac{863.13}{3} = 287.71$	0.5202	−1.6290

EXAMPLE 6.14 Find the fourth roots of −8.

SOLUTION The procedure is similar to that in Example 6.13. The polar coordinates of the point representing the given number are 8 and 180°. Letting r and θ denote the polar coordinates of the point representing a fourth root, we have

$$r^4\underline{/4\theta} = 8\underline{/360k° + 180°}$$

Then
$$r = 8^{1/4} = 1.6818$$

We now set k equal to 0, 1, 2, and 3 in turn and divide by 4 to obtain the values of θ. Converting from polar to rectangular form, we obtain the following as the fourth roots of -8:

$$1.1892(1 + i) \qquad 1.1892(-1 + i)$$

$$1.1892(-1 - i) \qquad 1.1892(1 - i)$$

CHAPTER 7

PLANE AND SOLID ANALYTIC GEOMETRY

7.1 PLANE ANALYTIC GEOMETRY

7.1.1 Definitions and Notation

In Art. 6.2.11, we defined the coordinates of a point that represents a complex number. In the present chapter, we shall define the coordinates of a point from a broader perspective.

In Fig. 7.1*a*, P is an arbitrary point in the plane, and QP and RP are lines through P perpendicular to the x and y axes, respectively. The distance OQ is the *x coordinate* or *abscissa* of P, and the distance OR is the *y coordinate* or *ordinate* of P. The x and y coordinates are known collectively as the *rectangular* (or *cartesian*) coordinates of P. The designation $P(x_1,y_1)$ identifies a point P that has an abscissa of x_1 and an ordinate of y_1.

If a displacement along a given line is considered to have an algebraic sign, the line is termed a *directed line*, and an arrowhead is used to identify the positive

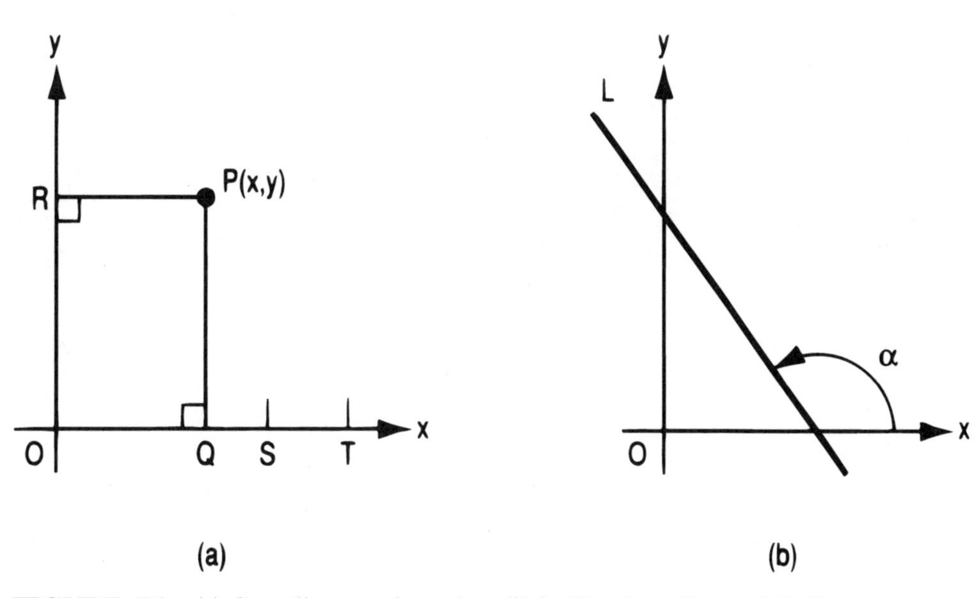

FIGURE 7.1 (*a*) Coordinates of a point; (*b*) inclination of a straight line.

direction. Thus, in Fig. 7.1*a*, the x axis is a directed line. The directed segment ST is positive, and the directed segment TS is negative. An *undirected line* is one to which no positive direction is assigned.

In Fig. 7.1*b*, L is an arbitrary undirected straight line. The *inclination* of L is the angle α between the positive side of the x axis and L, as measured in the counterclockwise direction. The *slope* of L is defined as the tangent of α.

If two curves, or a curve and a straight line, approach each other without ever intersecting, the lines are *asymptotes* of each other. As we proceed in a given direction, the distance between the two lines becomes and remains smaller than any number we can specify.

7.1.2 Equation of the Straight Line

In Fig. 7.2*a*, L is the straight line through points P_1 and P_2, which have the coordinates shown. Let m denote the slope of L. Then

$$m = \tan \alpha = \frac{y_2 - y_1}{x_2 - x_1} \tag{7.1}$$

In Fig. 7.2*b*, the straight line L contains point P_1 and intersects the x and y axes at $A(a,0)$ and $B(0,b)$, respectively. The distances a and b are called, respectively, the *x intercept* and *y intercept* of L. Again let m denote the slope of L. The equation for L can be given in any of the following forms:

$$y - y_1 = m(x - x_1) \tag{7.2}$$

$$y = mx + b \tag{7.3}$$

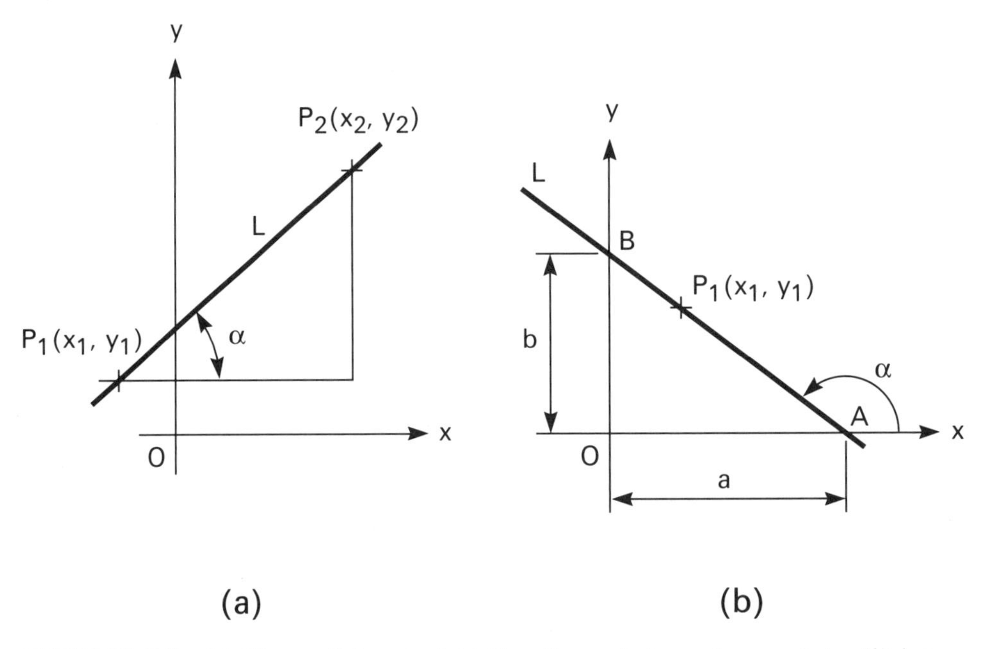

FIGURE 7.2 (*a*) Slope of the straight line through two given points; (*b*) intercepts of a straight line.

$$\frac{x}{a} + \frac{y}{b} = 1 \tag{7.4}$$

The *general form* of the equation of a straight line is

$$Ax + By + C = 0$$

where A, B, and C are real numbers. The slope of the line is

$$m = -\frac{A}{B} \tag{7.5}$$

In solving Examples 7.1 through 7.3, we recommend that the reader plot the given points as an aid in visualizing the problem. A crude sketch will suffice.

EXAMPLE 7.1 A straight line contains the points $P_1(-5,2)$ and $P_2(3,26)$. Write the equation of this line in general form. If $Q(x,20)$ lies on this line, what is x? Verify the results.

SOLUTION Applying Eqs. (7.1) and (7.2) in turn, we obtain

$$m = \frac{26 - 2}{3 - (-5)} = 3 \qquad y - 2 = 3[x - (-5)] = 3x + 15$$

or

$$3x - y + 17 = 0 \tag{a}$$

The coordinates of both P_1 and P_2 satisfy this equation, and it is thus confirmed.

Setting $y = 20$ in Eq. (*a*) and solving, we obtain $x = 1$, and that is the abscissa of Q. This result can be tested by applying Eq. (7.1) to find the slope of QP_2:

$$m = \frac{26 - 20}{3 - 1} = 3$$

The coordinates of Q satisfy the requirement.

EXAMPLE 7.2 A straight line has x and y intercepts of -6 and -4, respectively. Compute the angle that this line makes with the positive side of the x axis, and write the equation of the line in general form.

SOLUTION The line contains the points $A(-6,0)$ and $B(0,-4)$. By Eq. (7.1),

$$\tan \alpha = \frac{-4 - 0}{0 - (-6)} = -\frac{2}{3} \qquad \alpha = 146.31^\circ$$

By Eq. (7.4), the equation of the line is

$$\frac{x}{-6} + \frac{y}{-4} = 1 \qquad \text{or} \qquad 2x + 3y + 12 = 0$$

The coordinates of A and B satisfy this equation.

EXAMPLE 7.3 A straight line contains the point $P(20,9)$ and has a slope of 0.6. Locate the points at which this line intersects the x and y axes, and verify the results.

SOLUTION Again let $A(a,0)$ and $B(0,b)$ denote these points. Substituting the coordinates of P in Eq. (7.3), we obtain

$$9 = 0.6 \times 20 + b \qquad b = -3$$

Substituting the coordinates of P and A in Eq. (7.1), we obtain

$$0.6 = \frac{9-0}{20-a} \qquad a = 5$$

Thus, the line intersects the x axis at $A(5,0)$ and the y axis at $B(0,-3)$.

The results can be verified by applying Eq. (7.1) with the coordinates of A and B, giving

$$m = \frac{0-(-3)}{5-0} = 0.6$$

Alternatively, the results can be verified by applying Eq. (7.4) to obtain the equation of the line, which is $3x - 5y - 15 = 0$. The coordinates of P satisfy this equation.

EXAMPLE 7.4 A straight line has the equation $3x - 2y - 18 = 0$. Determine the slope of this line without applying Eq. (7.5).

SOLUTION Solving the equation for y, we obtain $y = 1.5x - 9$. Comparing this equation with Eq. (7.3), we find that $m = 1.5$.

7.1.3 Angle between Two Straight Lines

The acute angle between two straight lines can readily be found by determining the angle that each line makes with the x axis and then taking the difference between the results.

EXAMPLE 7.5 Lines L_1 and L_2 have the equations $4x + 9y - 36 = 0$ and $7x + 5y - 35 = 0$, respectively. Find the acute angle between these lines.

SOLUTION The lines are plotted in Fig. 7.3. Applying Eq. (7.5) or proceeding as in Example 7.4, we find that the slopes are as follows:

$$m_1 = -\frac{4}{9} \qquad m_2 = -\frac{7}{5}$$

Then

$$\alpha_1 = 156.04^\circ \qquad \alpha_2 = 125.54^\circ$$

As stated in Art. 6.1.4, an exterior angle of a triangle equals the sum of the two opposite interior angles. It follows that

$$\phi = \alpha_1 - \alpha_2 = 30.50^\circ$$

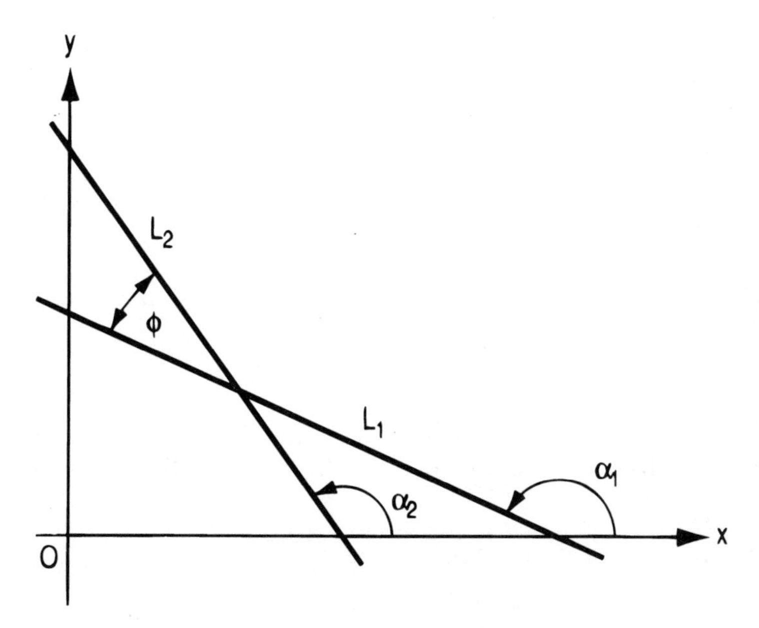

FIGURE 7.3 Angle between two straight lines.

Equation (6.13) yields this relationship: If the straight lines L_1 and L_2 have slopes of m_1 and m_2, the acute angle ϕ between these lines is

$$\phi = \arctan\left|\frac{m_1 - m_2}{1 + m_1 m_2}\right| \tag{7.6}$$

Since tan 90° is infinite, it follows that two straight lines are mutually perpendicular if

$$m_1 m_2 = -1 \tag{7.7}$$

EXAMPLE 7.6 Line L_1 has the equation

$$4x - 5y - 22 = 0 \tag{b}$$

Find the equation of the line L_2 that is perpendicular to L_1 and contains the point $P(21, -4)$.

SOLUTION Applying Eq. (7.5) or proceeding as in Example 7.4, we find that the slope of L_1 is 0.8. Therefore, by Eq. (7.7), the slope of L_2 is $-1/0.8 = -1.25$. Applying Eq. (7.2) with the coordinates of P, we obtain

$$y - (-4) = -1.25(x - 21)$$

Multiplying by 4 and rearranging, we now obtain

$$5x + 4y - 89 = 0 \tag{c}$$

as the equation of L_2.

EXAMPLE 7.7 With reference to Example 7.6, locate the point of intersection Q of lines L_1 and L_2.

SOLUTION Since Q lies on both lines, its coordinates satisfy Eqs. (b) and (c). The solution of this system of simultaneous equations is $x = 13$, $y = 6$. Therefore, the point of intersection of the lines is $Q(13,6)$.

7.1.4 Distance between Two Points

In accordance with the Pythagorean theorem, the distance d between points $P_1(x_1,y_1)$ and $P_2(x_2,y_2)$ is

$$d = \sqrt{(x_2 - x_1)^2 + (y_2 - y_1)^2} \tag{7.8}$$

EXAMPLE 7.8 Point $P(12,11)$ lies on a line that has a slope of 0.75. Locate the point Q on this line that lies at a distance of 9 from P and lies above P.

SOLUTION

Method 1. We shall first find the coordinates of Q algebraically. Let the subscripts 1 and 2 pertain to the coordinates of P and Q, respectively. Equation (7.1) yields

$$y_2 - 11 = 0.75(x_2 - 12) \tag{d}$$

Equation (7.8) then becomes

$$9 = \sqrt{1.5625(x_2 - 12)^2}$$

This equation yields two values: $x_2 = 19.2$ or 4.8; Eq. (d) then yields $y_2 = 16.4$ or 5.6, respectively. Since Q lies above P, its coordinates are $x = 19.2$ and $y = 16.4$.

Method 2. We shall now find the coordinates of Q trigonometrically. In Fig. 7.4, draw lines through P and Q parallel to the x and y axes, respectively, and label their intersection point R. Then $\tan \alpha = 0.75$. Since $PQ = 9$, we have the following:

$$PR = 9 \cos \alpha = 7.2 \qquad RQ = 9 \sin \alpha = 5.4$$

Therefore, the coordinates of Q are $x = 12 + 7.2 = 19.2$ and $y = 11 + 5.4 = 16.4$.

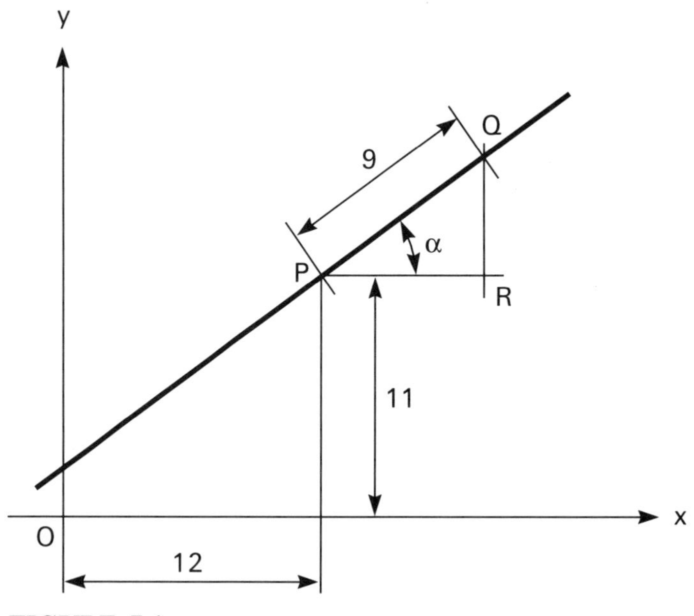

FIGURE 7.4

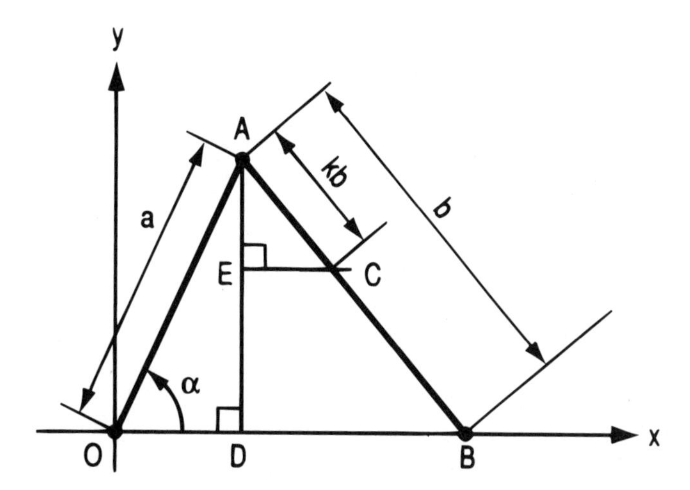

FIGURE 7.5 Motion of a point on a mechanism.

7.1.5 Parametric Equations

When a point moves in a prescribed manner, its instantaneous x and y coordinates may both be functions of a third variable, such as elapsed time or angular displacement. This third variable is termed a *parameter*, and the two equations that express the coordinates in terms of the parameter are called *parametric equations.*

EXAMPLE 7.9 In the mechanism in Fig. 7.5, the crank OA rotates about O, the rod AB is connected to OA with a smooth pin, and pin B is constrained to move along the x axis. The lengths are recorded in the drawing, and $b > a$. Express the coordinates of point C on AB in terms of the parameter α.

SOLUTION Draw DA parallel to the y axis and EC parallel to the x axis. The rectangular coordinates of C are as follows:

$$x = OD + EC = OD + k(DB) = OD + k\sqrt{(AB)^2 - (AD)^2}$$

$$y = DA - k(DA) = (1 - k)(DA)$$

Expressing all distances in term of a and b, we obtain

$$x = a \cos \alpha + k\sqrt{b^2 - a^2 \sin^2 \alpha}$$

$$y = a(1 - k) \sin \alpha$$

7.1.6 Polar Coordinates

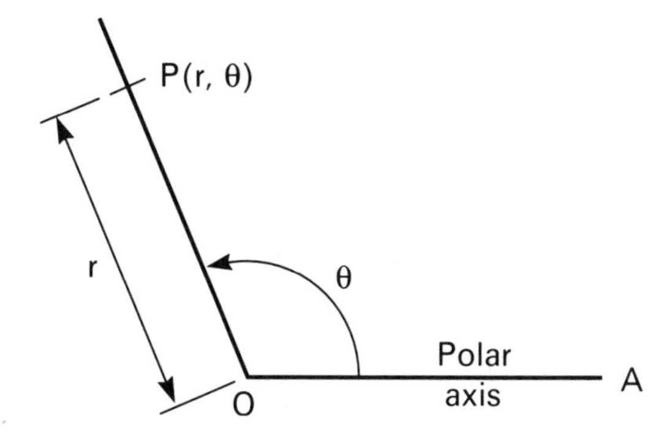

FIGURE 7.6 Polar coordinates of a point.

With reference to Fig. 7.6, the position of point P in the plane can be expressed by specifying the following: the distance of P from the reference point O, and the angle between the line OP and the reference line OA. Let $r = OP$ and $\theta = \angle AOP$, the angle being measured in the counterclockwise direction. The quantities r and θ are called, respectively, the *radius vector* and *vectorial angle* of P, and they constitute the *polar coordinates* of that point. Under this system, point O

is the *pole* and line OA is the *polar axis*. The notation $P(r_1,\theta_1)$ identifies a point P that has a radius vector r_1 and vectorial angle θ_1. It is often simpler to use polar coordinates rather than rectangular coordinates.

The two sets of coordinates are related in this manner:

$$r^2 = x^2 + y^2 \qquad \theta = \arctan \frac{y}{x} \tag{7.9}$$

$$x = r \cos \theta \qquad y = r \sin \theta \tag{7.10}$$

As previously stated, the conversion from one system of coordinates to another can be made by calculator.

When the equation of a curve is given in polar form, r is usually expressed as a function of θ.

EXAMPLE 7.10 When a circle of radius a has its center at $C(a,0)$, its equation in rectangular form is

$$x^2 + y^2 = 2ax$$

Formulate the corresponding equation in polar form.

SOLUTION

Method 1. Replacing x and y with their expressions as given by Eq. (7.10), we obtain

$$r^2 \cos^2 \theta + r^2 \sin^2 \theta = 2ar \cos \theta$$

or

$$r^2(\cos^2 \theta + \sin^2 \theta) = 2ar \cos \theta$$

Applying Eq. (6.5) and then dividing by r, we obtain

$$r = 2a \cos \theta$$

as the polar equation of this circle.

Method 2. The polar equation can be obtained directly without referring to the rectangular equation. The circle is shown in Fig. 7.7. Select an arbitrary point P on the curve and draw AP. Since $\overparen{OA}$ is a semicircle, $\angle OPA = 90°$. We have $OP = r$ and $\angle AOP = \theta$. Then $r = (OA) \cos \theta = 2a \cos \theta$.

7.1.7 Displacement of a Line

When a straight or curved line undergoes a displacement, the process is known as a *transformation*, and the equation of the line following the displacement is known as the *transformed equation*. It is often desirable to displace a line to obtain a simplified equation or for some other reason. We shall let $P(x,y)$ denote an arbitrary point on the line prior to the transformation and $P'(x',y')$ denote the corresponding point following the transformation.

There are three possible forms of displacement: a translation, a rotation, and a combination of the two. A straight line is translated when it moves in such manner

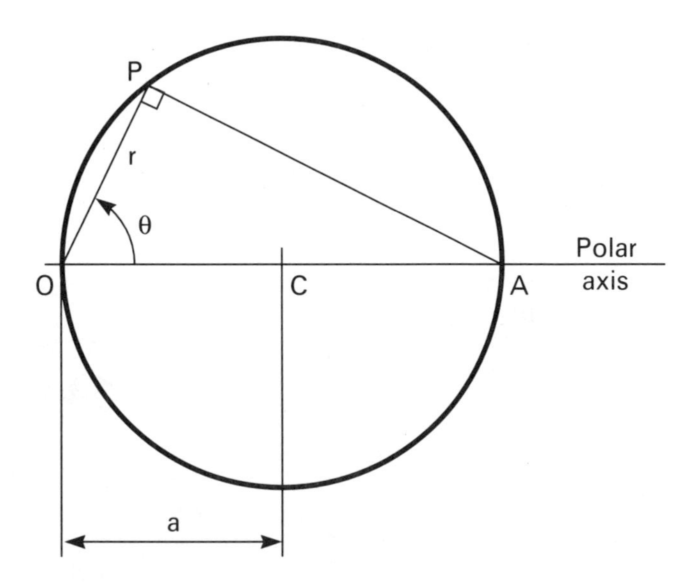

FIGURE 7.7 Formulating the polar equation of a curve.

that it remains parallel to its original position; a curve is translated when it moves in such manner that an arbitrary chord remains parallel to its original position.

Translation. Consider that the line is displaced a distance h in the positive x direction and a distance k in the positive y direction. The displacement of the arbitrary point P is shown in Fig. 7.8*a*, and the change in coordinates is as follows:

$$x' = x + h \qquad y' = y + k$$

Then

$$x = x' - h \qquad y = y' - k \tag{7.11}$$

The transformed equation of the line is obtained by replacing x and y in the original equation with their expressions in terms of x' and y' as given by Eq. (7.11) and then discarding the primes.

EXAMPLE 7.11 A curve has the equation

$$3x^2 - 2xy - 112 = 0$$

The curve is translated by moving it 4 units to the left and 9 units upward. Formulate the transformed equation of this curve, and test the result by selecting an arbitrary point on the curve.

SOLUTION In this instance, $h = -4$ and $k = 9$. Replacing x and y in the original equation with their expressions as given by Eq. (7.11), we obtain

$$3(x' + 4)^2 - 2(x' + 4)(y' - 9) - 112 = 0$$

Performing the required operations, rearranging, and then discarding the primes, we obtain the transformed equation

$$3x^2 - 2xy + 42x - 8y + 8 = 0$$

This result can be tested in the following manner: Select the point $P(8,5)$ that lies

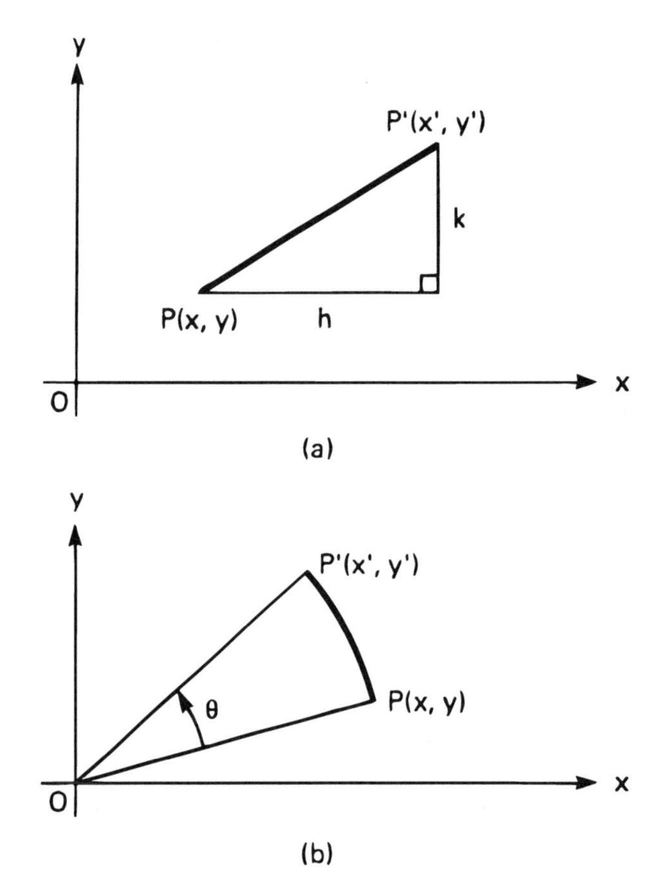

FIGURE 7.8 Displacement of a point (*a*) in translation and (*b*) in rotation.

on the curve in its original position. This point is transformed to $P'(4,14)$, and we find that the coordinates of P' satisfy the transformed equation.

Rotation. Consider that the line rotates about the origin in a counterclockwise direction through an angle θ. The displacement of the arbitrary point P is shown in Fig. 7.8*b*, and the coordinates are related in this manner:

$$x = x' \cos\theta + y' \sin\theta$$
$$y = -x' \sin\theta + y' \cos\theta \qquad (7.12)$$

In the special case where $\theta = 45°$, Eqs. (7.12) reduce to the following:

$$x = \frac{x' + y'}{\sqrt{2}} \qquad y = \frac{-x' + y'}{\sqrt{2}} \qquad (7.12a)$$

In the special case where $\theta = 90°$, Eqs. (7.12) reduce to the following:

$$x = y' \qquad y = -x' \qquad (7.12b)$$

EXAMPLE 7.12 A straight line has the equation $5x - 12y + 76 = 0$. The line rotates about the origin in a counterclockwise direction through an angle of 90°. Formulate the transformed equation of the line, and test it for compliance with Eq. (7.7).

SOLUTION Replacing x and y with their expressions as given by Eq. (7.12*b*) and discarding the primes, we obtain $5y + 12x + 76 = 0$, or $12x + 5y + 76 = 0$.

By Eq. (7.5), the slope of the line is 5/12 in the original position and $-12/5$ in the transformed position. Thus, Eq. (7.7) is satisfied.

7.2 THE CONIC SECTIONS

7.2.1 Definitions and Types of Conics

The line of intersection of a plane and a right circular cone is termed a *conic section*, or simply a *conic*. However, a conic section can be defined and analyzed merely by studying the motion of a point in a plane, and that is the conventional method that is used.

In Fig. 7.9*a*, F is a fixed point, L is a fixed line, P is a moving point, and PR and FH are perpendicular to L. Point F is the *focus*, line L is the *directrix*, line FP is the *focal radius*, and FH is the *principal axis*. The directrix may lie to the right of the focus, as shown, or to the left. The distances FP and PR are considered to be

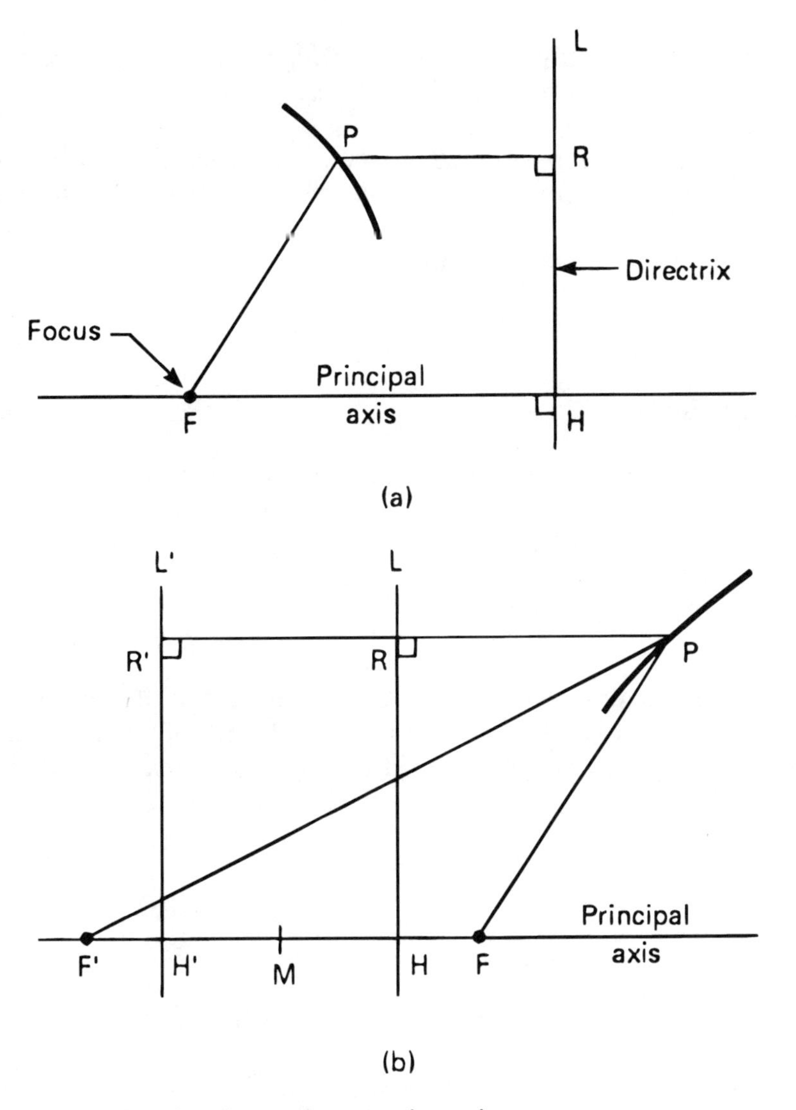

FIGURE 7.9 Generating a conic section.

positive. We define the *eccentricity* e in this manner:

$$e = \frac{FP}{PR} \tag{7.13}$$

If point P moves in such manner that e remains constant, the curve generated by P is a conic section.

The type of section formed is as follows: an *ellipse* if $e < 1$, a *parabola* if $e = 1$, and a *hyperbola* if $e > 1$. The circle is regarded as a degenerate form of the ellipse in which the directrix is at an infinite distance from the focus and therefore $e = 0$. The ellipse is a closed curve, the parabola is an open curve, and the hyperbola is an open curve that has two symmetrical branches.

As shown in Fig. 7.9*b*, some conic sections have two foci, F and F', and two directrices, L and L'. The directrices are perpendicular to the line through the foci, and $H'F' = FH$. Line $H'H$ is the principal axis. Point M, which is the midpoint of $H'H$ and of $F'F$, is called the *center* of the conic. In this situation,

$$\frac{FP}{PR} = \frac{F'P}{R'P} = e$$

The chord of a conic section that passes through a focus and is perpendicular to the principal axis is the *latus rectum*. The point at which the principal axis intersects the conic is the *vertex*.

When a conic section is positioned in such a manner that its equation in rectangular coordinates is as simple as possible, the section is said to be in *standard position*, and the corresponding equation is termed its *standard form*. We shall exhibit each conic section in standard position.

7.2.2 The Parabola

As previously stated, the moving point P in Fig. 7.9*a* generates a parabola if $e = 1$; i.e., if $FP = PR$. In Fig. 7.10, we have placed the principal axis of the parabola on the x axis and its vertex at the origin. Let $AF = p$. Since O lies on the parabola, $AO = OF = p/2$. The equation of the parabola is

$$y^2 = 2px \tag{7.14}$$

The curve is symmetrical about the x axis. The chord $K'K$ is the latus rectum. By setting $x = p/2$, we find that $FK = p$; therefore, the length of the latus rectum is $2p$.

Consider that the parabola is translated to place its vertex at (h, k). Applying Eq. (7.11) and discarding the primes, we obtain the transformed equation

$$(y - k)^2 = 2p(x - h) \tag{7.14a}$$

EXAMPLE 7.13 A parabola has the equation

$$8x - 12y - y^2 - 108 = 0 \tag{e}$$

Determine the following: the location of its focus; the location of its directrix; the length of its latus rectum. Verify the locations of the focus and directrix.

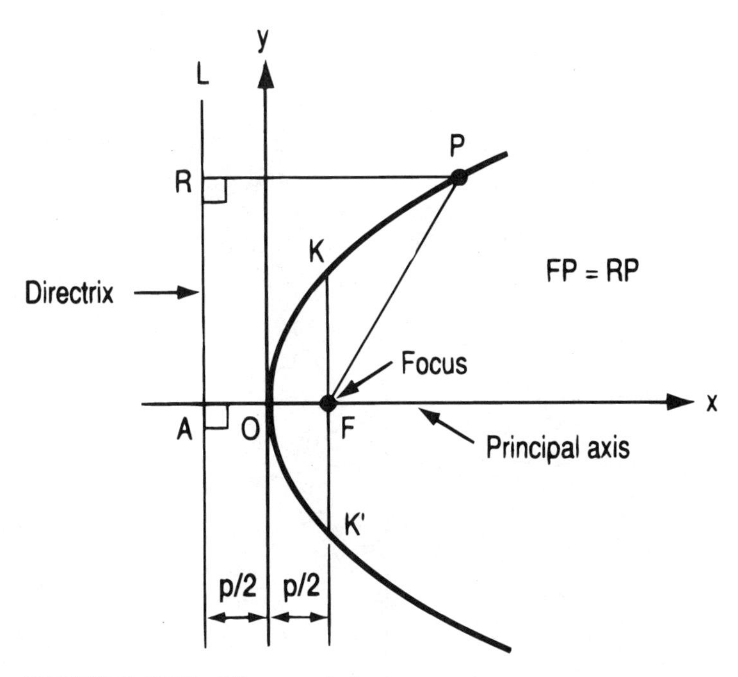

FIGURE 7.10 The parabola.

SOLUTION Equation (7.14*a*) can be rewritten in this form:

$$2px + 2ky - y^2 - (k^2 + 2ph) = 0 \tag{7.14b}$$

Comparing Eq. (*e*) with Eq. (7.14*b*), we obtain these values:

$$2p = 8 \quad \therefore p = 4 \qquad 2k = -12 \quad \therefore k = -6$$

$$k^2 + 2ph = 108 \quad \therefore h = 9$$

Thus, the vertex of the curve lies at $(9, -6)$. From Fig. 7.10, we find that the abscissa of the focus is $9 + p/2 = 11$, and the abscissa of A is $9 - p/2 = 7$. Therefore, the focus lies at $F(11, -6)$ and the directrix lies at a distance of 7 to the right of the y axis. (Alternatively, we may say that the directrix is the line $x = 7$.) The length of the latus rectum is $2p$, or 8.

The locations of the focus and directrix are verified in this manner: Let $P(x,y)$ denote an arbitrary point on the parabola. Let m denote the distance from P to the focus and n denote the distance from P to the directrix. Then

$$m^2 = (x - 11)^2 + (y + 6)^2 \qquad n^2 = (x - 7)^2$$

By expanding all terms, subtracting m^2 from n^2, and comparing the difference with Eq. (*e*), we find that $m = n$. Our results are thus confirmed.

Now consider that the parabola is initially in the position shown in Fig. 7.10 and then rotates about the origin in the counterclockwise direction through an angle of 90°. Applying the expressions in Eq. (7.12*b*) to Eq. (7.14) and discarding the primes, we obtain the transformed equation

$$y = \frac{x^2}{2p} \tag{7.14c}$$

7.2.3 The Ellipse

The ellipse has two foci, F and F', and two directrices, L and L'. In Fig. 7.11, we have placed the principal axis of the ellipse on the x axis and its center at the origin. Set $OF = c$ and $a = c/e$, where e again denotes the eccentricity. Since $e < 1$, $a > c$. Now let $b^2 = a^2 - c^2$. Figure 7.12 shows the location of the directrices. If we place each directrix at a distance of a/e from the y axis, the equation of the ellipse

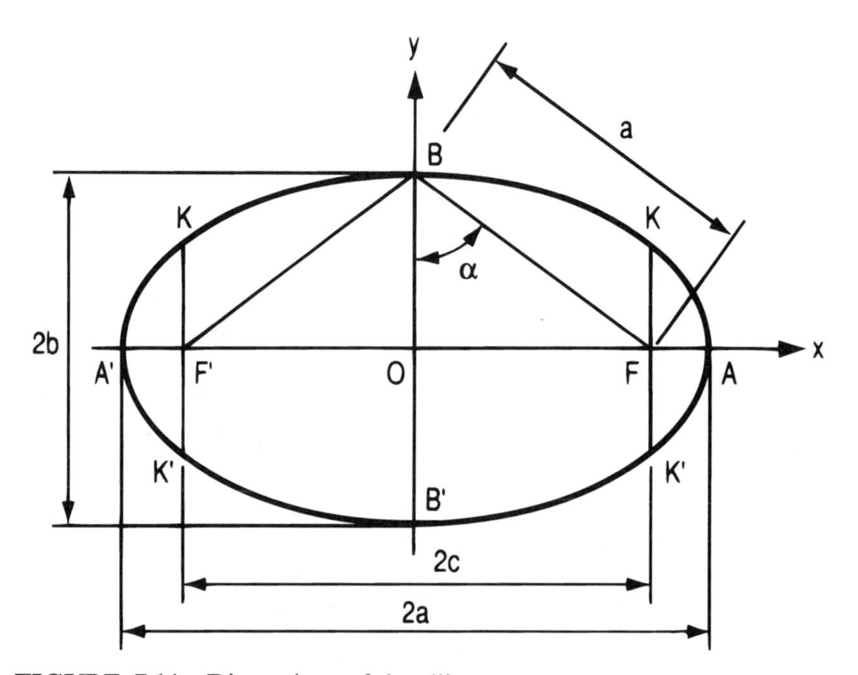

FIGURE 7.11 Dimensions of the ellipse.

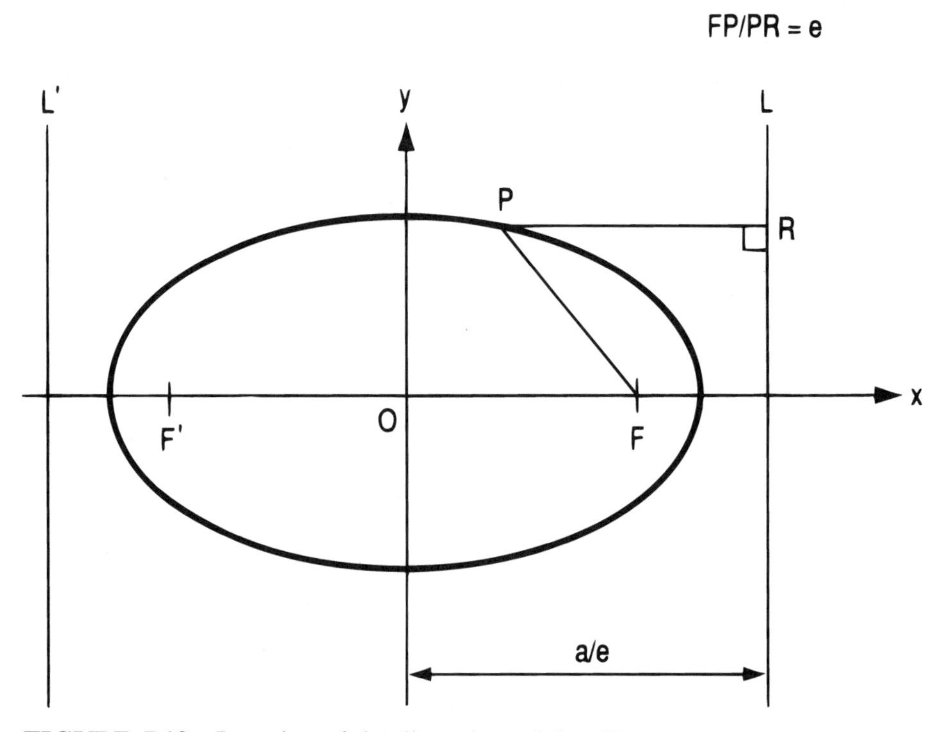

FIGURE 7.12 Location of the directrices of the ellipse.

becomes

$$\frac{x^2}{a^2} + \frac{y^2}{b^2} = 1 \tag{7.15}$$

This equation yields the following dimensions in Fig. 7.11: $OA = a$, $OB = b$. From the definition of b, we have $FB = a$. Then

$$e = \frac{c}{a} = \sin \alpha \tag{7.16}$$

Thus, the greater the eccentricity, the flatter the curve. By setting $x = c$, we find that the length of the latus rectum KK' in Fig. 7.11 is $2b^2/a$.

If the ellipse is translated to place its center at (h,k), the transformed equation is

$$\frac{(x-h)^2}{a^2} + \frac{(y-k)^2}{b^2} = 1 \tag{7.15a}$$

Let P denote an arbitrary point on the curve. The ellipse has the following highly significant property:

$$F'P + FP = 2a \tag{7.17}$$

Therefore, the ellipse can be defined alternatively as the trajectory of a point that moves in such manner than the sum of its distances from two fixed points remains constant.

EXAMPLE 7.14 An ellipse has the equation $25x^2 + 169y^2 = 4225$. Compute the eccentricity of this ellipse. Find the ordinate of a point on the curve that lies on a line parallel to the y axis and midway between that axis and the latus rectum.

SOLUTION By dividing both sides of the given equation by 4225, we obtain the standard form

$$\frac{x^2}{169} + \frac{y^2}{25} = 1$$

Therefore, $a = 13$, $b = 5$, and $c = 12$. Then $e = c/a = 0.9231$. Now refer to Fig. 7.11 and set $x = c/2 = 6$. The given equation yields $y = \pm 4.4356$.

EXAMPLE 7.15 In Fig. 7.13, the rod AB (called a *trammel*) has marks at the fixed points C, D, and E. The rod is displaced in such manner that C is constrained to move along the y axis and D is constrained to move along the x axis. Identify the type of curve that is generated by point E.

SOLUTION Draw FE parallel to the y axis and CF parallel to the x axis. Let $m = CE$ and $n = DE$, and let θ denote the angle that the rod makes with the x axis, as shown. The coordinates of E are $x = OG = CF$ and $y = GE$. Then

$$\frac{CF}{CE} = \frac{x}{m} = \cos \theta \qquad \frac{GE}{DE} = \frac{y}{n} = \sin \theta$$

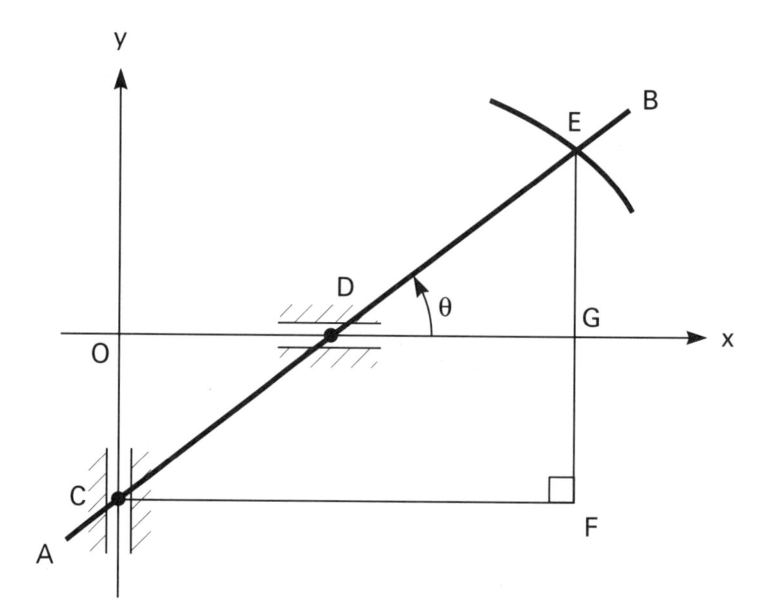

FIGURE 7.13 Trammel method of generating an ellipse.

By Eq. (6.5),

$$\frac{x^2}{m^2} + \frac{y^2}{n^2} = 1$$

Thus, the curve generated by point E is an ellipse. Since $m > n$, the curve has its major axis on the x axis and its minor axis on the y axis.

This method of generating an ellipse is widely applied, and it is called the *trammel method.* The ellipsograph, which is a device that generates an ellipse, is based on this method.

7.2.4 The Circle

This conic is a degenerate form of the ellipse in which the foci are coincident and the directrices are at an infinite distance from the center. Let r denote the radius of the circle. If the center of the circle is at the origin, its equation is simply

$$x^2 + y^2 = r^2 \tag{7.18}$$

If the center is at (h,k), the equation is

$$(x - h)^2 + (y - k)^2 = r^2 \tag{7.18a}$$

The circle may be viewed as the limiting shape approached by an ellipse as b increases and approaches a.

7.2.5 The Hyperbola

The hyperbola also has two foci, F and F', and two directrices, L and L'. In Fig. 7.14, we have placed the principal axis of the hyperbola on the x axis and its center at the origin. As before, let $OF = c$ and $a = c/e$, where e again denotes the eccentricity. Since $e > 1$, $a < c$. Now let $b^2 = c^2 - a^2$. Figure 7.15 shows the location of a direc-

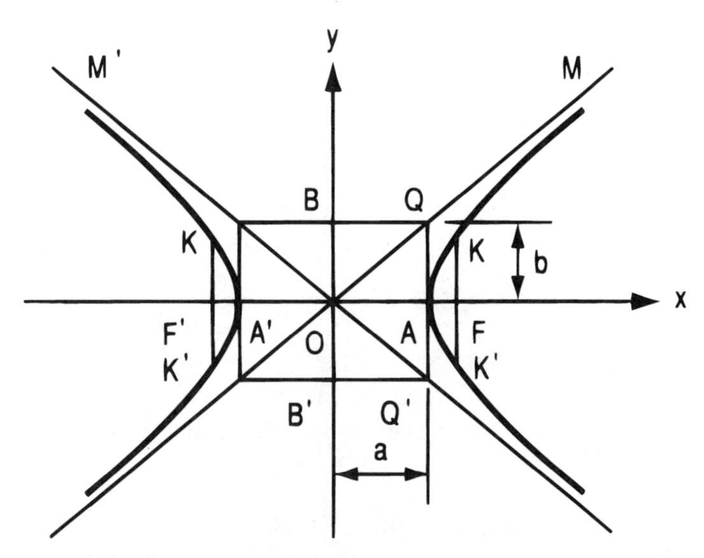

FIGURE 7.14 Axes and asymptotes of the hyperbola.

trix. If we place this at a distance of a/e from the y axis, the equation of the hyperbola becomes

$$\frac{x^2}{a^2} - \frac{y^2}{b^2} = 1 \tag{7.19}$$

The curve has two branches, and it is symmetrical about both the x and y axes.

In Fig. 7.14, lines M and M' have the equations $y = bx/a$ and $y = -bx/a$, respectively. Equation (7.19) reveals that the ratio y/x approaches the limit b/a or $-b/a$ as x becomes infinite. Therefore, lines M and M' are asymptotes of the hyperbola.

From Eq. (7.19), $OA = a$. Line $Q'Q$ is parallel to the y axis and contains A. Since Q lies on M, $AQ = b$. We construct the rectangle shown. Lines $A'A$ and $B'B$ are the *transverse axis* and *conjugate axis*, respectively, of the hyperbola. By setting $x = c$, we find that the length of the latus rectum $K'K$ in Fig. 7.14 is $2b^2/a$.

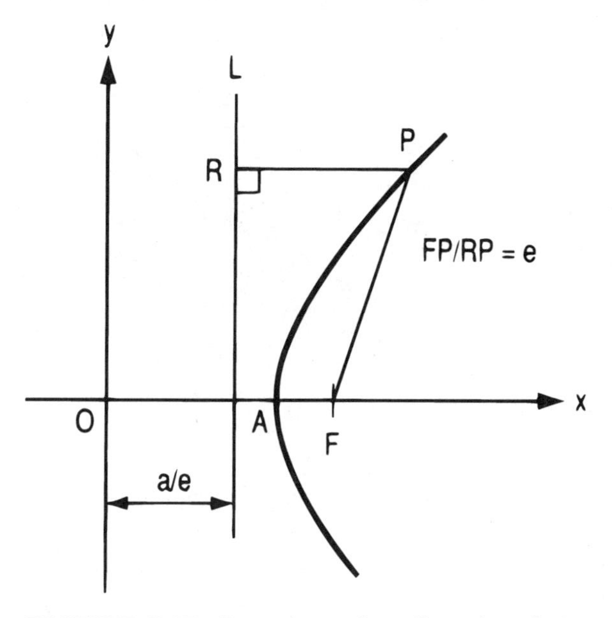

FIGURE 7.15 Location of a directrix of the hyperbola.

If the hyperbola is translated to place its center at (h,k), its equation becomes

$$\frac{(x-h)^2}{a^2} - \frac{(y-k)^2}{b^2} = 1 \tag{7.19a}$$

Let P denote an arbitrary point on the curve. The hyperbola has the following highly significant property:

$$F'P - FP = \pm 2a \tag{7.20}$$

We apply the plus or minus sign, respectively, according to whether P lies to the right or to the left of the y axis. Therefore, the hyperbola can be defined alternatively as the trajectory of a point that moves in such manner that the difference between its distances from two fixed points remains constant.

EXAMPLE 7.16 A hyperbola in standard position contains the point $P(26,-12)$, and its asymptotes have the equation $y = \pm 0.5x$. Formulate the equation of the curve, and find its eccentricity.

SOLUTION Substituting the given coordinates in Eq. (7.19), we obtain

$$\frac{676}{a^2} - \frac{144}{b^2} = 1 \tag{f}$$

From the general equations for lines M and M' in Fig. 7.14, we obtain $b/a = 0.5$; then $a = 2b$. Replacing a with $2b$ in Eq. (f) and solving, we obtain $b = 5$; then $a = 10$. Substituting these values in Eq. (7.19) and then multiplying by 100, we obtain $x^2 - 4y^2 = 100$ as the equation of the curve.

From the definitions of a and b, we have

$$e = \frac{c}{a} = \frac{\sqrt{a^2 + b^2}}{a} = \frac{\sqrt{125}}{10} = 1.1180$$

An *equilateral hyperbola* is one having the property $a = b$. Thus, its eccentricity is $\sqrt{2}$. When the equilateral hyperbola is in standard position, its equation is

$$x^2 - y^2 = a^2 \tag{7.21}$$

The asymptotes make 45° angles with the x and y axes. Now consider that the equilateral hyperbola rotates about the origin in the counterclockwise direction through an angle of 45°. Applying Eq. (7.12a) and discarding the primes, we obtain the transformed equation

$$xy = \frac{a^2}{2} \tag{7.22}$$

The x and y axes are the asymptotes.

7.2.6 General Equation of the Conic Section

When a conic section is placed in a nonstandard position, its equation has the following general form:

$$Ax^2 + 2Bxy + Cy^2 + 2Dx + 2Ey + F = 0$$

where the coefficients are constants and A or C, or both, have a nonzero value. Conversely, a second-degree equation that can be plotted represents a conic. The nature of the conic can be ascertained by computing this quantity:

$$J = B^2 - AC$$

The conic is an ellipse if $J < 0$, a parabola if $J = 0$, and a hyperbola if $J > 0$. The circle is a special form of conic in which $A = C$ and $B = 0$. Thus, the nature of the conic is determined solely by the first three terms in the general equation.

EXAMPLE 7.17 Identify the type of conic represented by each of the following equations:

$$8x^2 + 10xy + 3y^2 - 7x - 58 = 0 \qquad (g)$$

$$3x^2 - 8xy + 7y^2 - 11x + 13y + 6 = 0 \qquad (h)$$

$$5x^2 - 6xy - 9y^2 + 18y - 23 = 0 \qquad (i)$$

SOLUTION

Equation (g) $\quad B = 10/2 = 5 \qquad J = 5^2 - 8 \times 3 = 1$

The curve is a hyperbola.

Equation (h) $\quad B = -8/2 = -4 \qquad J = (-4)^2 - 3 \times 7 = -5$

The curve is an ellipse.

Equation (i) $\quad B = -6/2 = -3 \qquad J = (-3)^2 - 5(-9) = 51$

The curve is a hyperbola.

7.3 SOLID ANALYTIC GEOMETRY

7.3.1 Rectangular Coordinates of a Point

The x, y, and z coordinate axes in Fig. 7.16 are mutually perpendicular, they intersect at O (the *origin*), and they have the indicated positive directions. A plane that contains two coordinate axes is a *coordinate plane*, and it is labeled to indicate which axes it contains. For example, the xz plane contains the x and z axes.

Let P denote a point in space. The *rectangular* (or *cartesian*) *coordinates* of P are found by passing planes through P parallel to the coordinate planes, thereby forming the rectangular block in Fig. 7.16. Then $x = OA$, $y = OB$, $z = OC$. The designation $P(x_1,y_1,z_1)$ denotes a point having the specified coordinates.

7.3.2 Distance between Two Points

Let d denote the distance between points $P_1(x_1,y_1,z_1)$ and $P_2(x_2,y_2,z_2)$. By applying the Pythagorean theorem twice, we obtain

$$d = \sqrt{(x_2 - x_1)^2 + (y_2 - y_1)^2 + (z_2 - z_1)^2} \qquad (7.23)$$

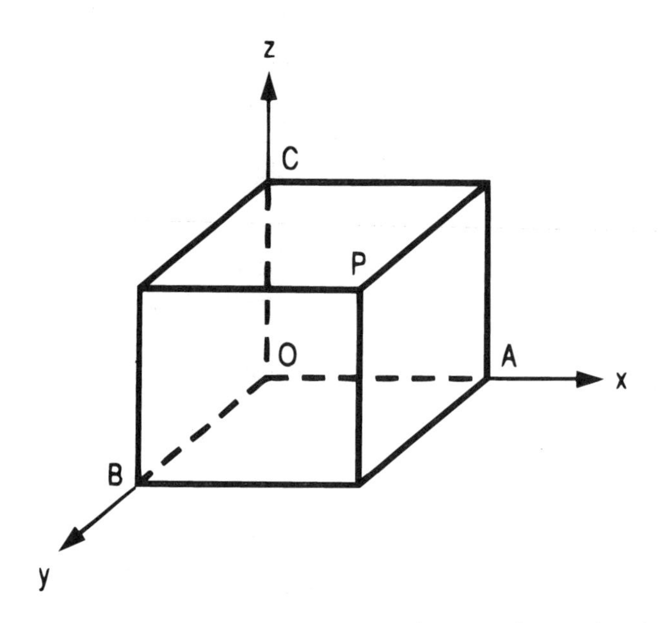

FIGURE 7.16 Rectangular coordinates of a point in three-dimensional space.

EXAMPLE 7.18 Compute the distance between the points $P_1(-4,3,9)$ and $P_2(10,-5,-7)$.

SOLUTION

$$d^2 = [10-(-4)]^2 + (-5-3)^2 + (-7-9)^2 = 516 \qquad d = 22.72$$

EXAMPLE 7.19 With reference to Example 7.18, find the length of line P_1P_2 as projected onto the xy plane and the yz plane.

SOLUTION The length of a line in space as projected onto the xy plane is governed solely by its lengths as projected onto the x and y axes. Let d_{xy} denote this length. For the given line, Eq. (7.8) yields

$$d_{xy}^2 = [10-(-4)]^2 + (-5-3)^2 = 260 \qquad d_{xy} = 16.12$$

$$d_{yz}^2 = (-5-3)^2 + (-7-9)^2 = 320 \qquad d_{yz} = 17.89$$

7.3.3 Direction Cosines and Direction Components

Let L denote a directed straight line that passes through the origin. The angle between the positive side of a coordinate axis and the positive direction of L is a *direction angle* of L, and the cosine of this angle is a *direction cosine* of L. These definitions are illustrated in Fig. 7.17, where P is an arbitrary point on L and α, β, and γ are the direction angles of L with reference to the x, y, and z axes, respectively. The direction cosines of L and their designations are as follows:

$$\cos\alpha = l = \frac{OA}{OP} \qquad \cos\beta = m = \frac{OB}{OP} \qquad \cos\gamma = n = \frac{OC}{OP} \tag{7.24}$$

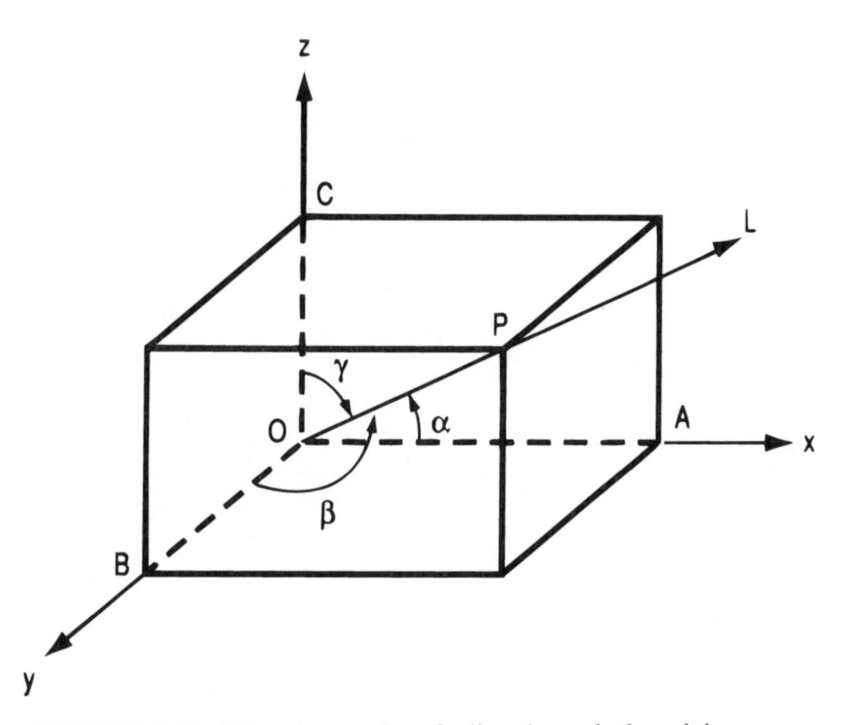

FIGURE 7.17 Direction angles of a line through the origin.

By Eq. (7.23),

$$(OP)^2 = (OA)^2 + (OB)^2 + (OC)^2$$

and it follows that

$$l^2 + m^2 + n^2 = 1 \tag{7.25}$$

EXAMPLE 7.20 Line L passes through the origin, and it has the direction angles $\alpha = 41°$ and $\beta = 118°$. Point P lies on L and above the xy plane; OP has the positive direction of L, and $OP = 15$. Find the following: the coordinates of P; the angle δ between the positive side of the z axis and the projection of OP onto the yz plane.

SOLUTION Since β is obtuse, line OP lies on the far side of the xz plane. The x and y coordinates of P are

$$x = 15 \cos 41° = 11.32 \qquad y = 15 \cos 118° = -7.04$$

Since P lies above the xy plane, its z coordinate is positive, and its coordinates are related in this manner:

$$x^2 + y^2 + z^2 = (OP)^2$$

Then $$z^2 = 15^2 - [(11.32)^2 + (-7.04)^2] \qquad z = 6.88$$

Refer to Fig. 7.18. In this drawing, the observer's line of sight is normal to the yz plane, and line OP' is the projection of OP onto this plane.

$$\tan \delta = \frac{7.04}{6.88} \qquad \delta = 45.66°$$

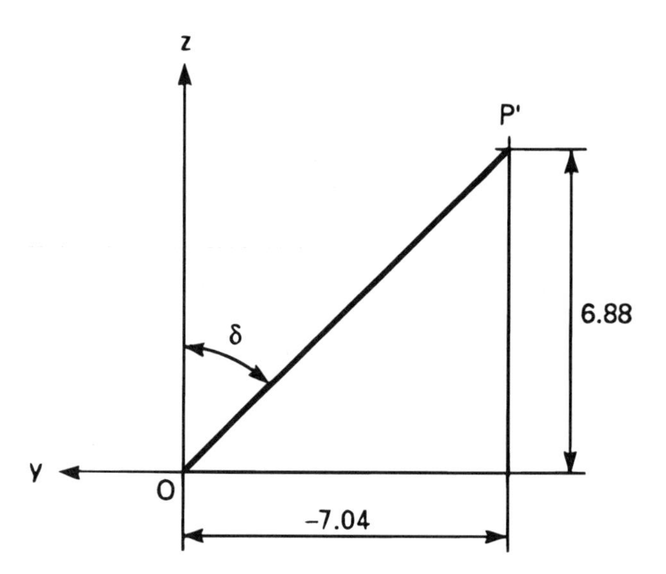

FIGURE 7.18 Projection of OP onto the xy plane.

If line L does not pass through the origin, we draw a line L' through the origin parallel to L and having the same positive direction as L. The direction angles of L' are also those of L.

Let $P(x_1,y_1,z_1)$ and $P_2(x_2,y_2,z_2)$ denote arbitrary points on the directed line L, where P_1P_2 has the positive direction of the line. Then

$$l = \frac{x_2 - x_1}{P_1P_2} \qquad m = \frac{y_2 - y_1}{P_1P_2} \qquad n = \frac{z_2 - z_1}{P_1P_2} \tag{7.26}$$

EXAMPLE 7.21 A line contains the points $P_1(8,7,3)$ and $P_2(11,19,7)$, where P_1P_2 has the positive direction of the line. Compute the direction angles of this line.

SOLUTION Equation (7.23) yields $P_1P_2 = 13$. Then

$$l = \frac{11-8}{13} = \frac{3}{13} \qquad m = \frac{19-7}{13} = \frac{12}{13} \qquad n = \frac{7-3}{13} = \frac{4}{13}$$

$$\alpha = \arccos\,(3/13) = 76.66^\circ \qquad \beta = \arccos\,(12/13) = 22.62^\circ$$

$$\gamma = \arccos\,(4/13) = 72.08^\circ$$

Now consider that the direction cosines l, m, and n are multiplied by some constant, and let a, b, and c denote the respective products. Then a, b, and c constitute a set of *direction components* of L. The direction components provide a convenient means of visualizing the direction of a line.

EXAMPLE 7.22 A line has the direction components 3, 8, and -5 with reference to the x, y, and z axes, respectively. As we move from point P to point Q along this line, x increases by 12 units. Find the corresponding change in the y and z coordinates.

SOLUTION The direction components inform us that if x increases by 3 units, y increases by 8 units and z decreases by 5 units. Therefore, when x increases by 12 units, y increases by $4 \times 8 = 32$ units and z decreases by $4 \times 5 = 20$ units.

7.3.4 General Equation of a Plane

The general equation of a plane has the form $Ax + By + Cz + D = 0$.

EXAMPLE 7.23 A plane has the equation $6x - 5y + 12z - 90 = 0$. Locate the points at which this plane intersects the coordinate axes.

SOLUTION Setting $y = z = 0$, we obtain $x = 15$. Setting $x = z = 0$, we obtain $y = -18$. Setting $x = y = 0$, we obtain $z = 7.5$. Therefore, the plane intersects the x axis at (15,0,0), the y axis at (0,−18,0), and the z axis at (0,0,7.5). These points of intersection are the *intercepts* of the plane.

7.3.5 Equation of a Straight Line

The equation of a straight line in space can be given in three alternative forms, and the form to be applied in a specific situation is determined by the given information. Let P denote a point on the line. If we are given one coordinate of P, the equation of the line must enable us to determine the two remaining coordinates of P. It follows that the equation of the line is a composite of *two equations.*

In the *general form* of the composite equation, we visualize that the line results from the intersection of two planes, and we record the equations of these planes.

EXAMPLE 7.24 A straight line has the composite equation

$$6x + 7y + 12z - 62 = 0$$

$$3x - 4y + 10z + 17 = 0$$

Point P on this line has an x coordinate of -5. Find the y and z coordinates of P.

SOLUTION When we replace x in the given equation with -5, we obtain the following equations:

$$7y + 12z = 92$$

$$-4y + 10z = -2$$

The solution of this system of simultaneous equations is $y = 8$ and $z = 3$, and these are the remaining coordinates of P.

Now let a, b, and c denote, respectively, x, y, and z direction components of the line, and let $P(x_1,y_1,z_1)$ denote a point that lies on this line. As previously stated, the direction components reveal the relative amounts by which the coordinates increase as we move along the line. It follows that

$$\frac{x - x_1}{a} = \frac{y - y_1}{b} = \frac{z - z_1}{c} \tag{7.27}$$

This is the *symmetric form* of the composite equation of the line.

EXAMPLE 7.25 A straight line has direction components of 5, 3, and -6 with reference to the x, y, and z axes, respectively, and it contains the point $P(4,-2,9)$. Write the equation of the line, and find the x and y coordinates of the point Q where $z = 21$.

SOLUTION Equation (7.27) yields

$$\frac{x-4}{5} = \frac{y+2}{3} = \frac{z-9}{-6}$$

as the equation of the line. When $z = 21$, the third fraction is -2. Therefore, the remaining coordinates of Q are $x = -6$ and $y = -8$.

As we move from Q to P, x increases by 10, y increases by 6, and z decreases by 12. These changes are consistent with the direction components of the line.

Now consider that a line contains the points $P_1(x_1,y_1,z_1)$ and $P_2(x_2,y_2,z_2)$. As we move along the line, the coordinates change in fixed proportions, and it follows that

$$\frac{x-x_1}{x_2-x_1} = \frac{y-y_1}{y_2-y_1} = \frac{z-z_1}{z_2-z_1} \tag{7.28}$$

This is the *two-point form* of the composite equation of the line.

7.3.6 Cylindrical and Spherical Coordinates

A point in space can be located by means of the rectangular coordinate system described in Art. 7.3.1, but two other coordinate systems are available. To describe these systems, we shall let P denote a point having the rectangular coordinates x, y, and z, and we shall let P' denote the projection of P onto the xy plane (i.e., the foot of the perpendicular from P to that plane).

In Fig. 7.19*a*, let r and θ denote the polar coordinates of P' as defined in Art. 7.1.6. The quantities r, θ, and z constitute the *cylindrical coordinates* of P, and the point is identified by the notation $P(r,\theta,z)$.

In Fig. 7.19*b*, let $OP = \rho$, let θ denote the angle from the positive side of the x axis to OP' as measured in the direction shown, and let ϕ denote the angle from the

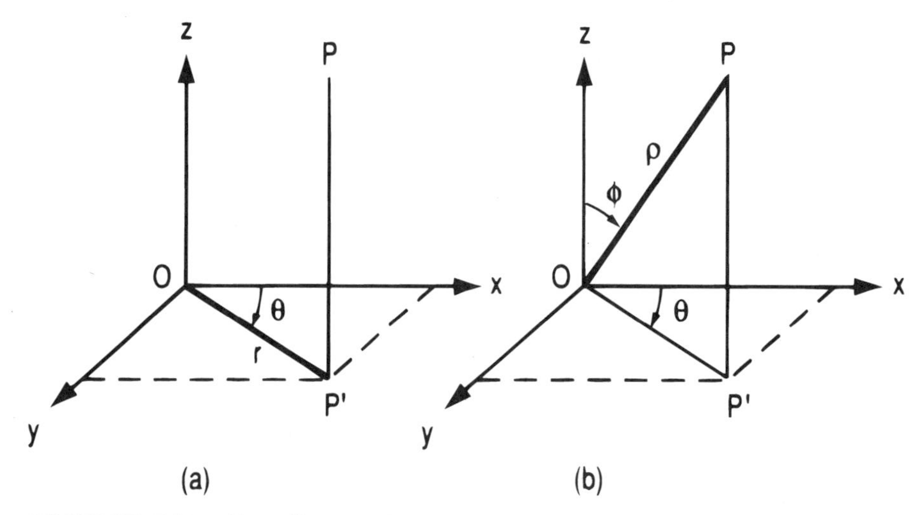

FIGURE 7.19 Coordinates of a point in space. (*a*) Cylindrical coordinates; (*b*) spherical coordinates.

positive side of the z axis to OP. The quantities ρ, θ, and ϕ constitute the *spherical coordinates* of P, and the point is identified by the notation $P(\rho,\theta,\phi)$. We have $OP' = \rho \sin \phi$. The spherical coordinates can be transformed to the rectangular coordinates by these relationships:

$$x = \rho \sin \phi \cos \theta$$

$$y = \rho \sin \phi \sin \theta \qquad (7.29)$$

$$z = \rho \cos \phi$$

We shall now illustrate how rectangular coordinates are transformed to spherical coordinates. In doing so, we shall apply the following relationship, which stems from Eq. (7.23):

$$\rho = \sqrt{x^2 + y^2 + z^2} \qquad (7.30)$$

EXAMPLE 7.26 Point P has the rectangular coordinates $x = -6$, $y = -9$, $z = 7$. Find the corresponding spherical coordinates.

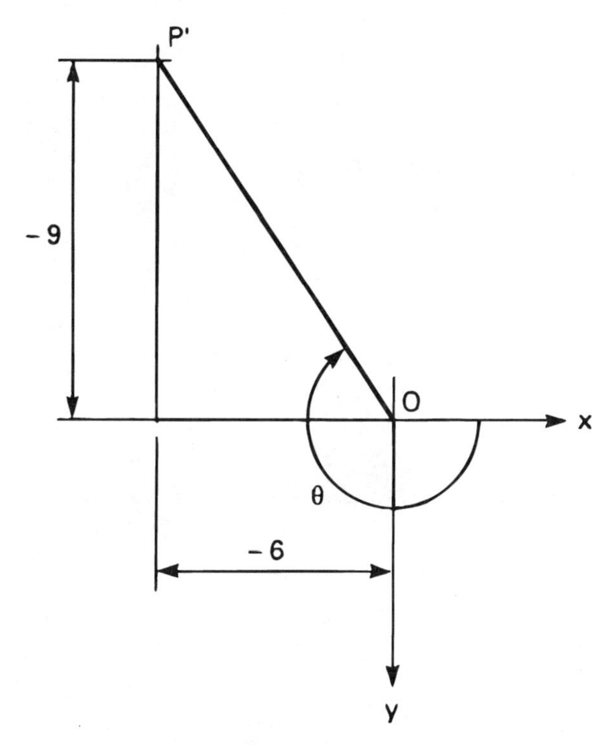

FIGURE 7.20 Location of P' in xy plane.

SOLUTION Refer to Fig. 7.20, which shows the location of P' in the xy plane.

$$\tan \theta = \frac{-9}{-6} = 1.5 \qquad \theta = 56.31° + 180° = 236.31°$$

$$OP' = \sqrt{(-6)^2 + (-9)^2} = 10.8167$$

$$\tan \phi = \frac{OP'}{z} = \frac{10.8167}{7} \qquad \phi = 57.09°$$

$$\rho = \sqrt{(-6)^2 + (-9)^2 + 7^2} = 12.8841$$

Thus, the spherical coordinates are $\rho = 12.8841$, $\theta = 236.31°$, $\phi = 57.09°$. When we substitute these values in Eqs. (7.29), we obtain the given rectangular coordinates.

CHAPTER 8

DIFFERENTIAL CALCULUS

8.1 METHODS OF DIFFERENTIATION

8.1.1 Definition and Significance of the Derivative

Let y be a function of x. Expressed symbolically, $y = f(x)$. Consider that we assign a value to x, thereby establishing the value of y. Now consider that we increment x by an amount Δx, and let Δy denote the corresponding increment of y. The *derivative* of y with respect to x, which is denoted by dy/dx, is the following:

$$\frac{dy}{dx} = \lim_{\Delta x \to 0} \frac{\Delta y}{\Delta x} \tag{8.1}$$

Alternatively, the derivative may be denoted by y', $f'(x)$, $D_x y$, or $\dot{y}$.

The geometric significance of the derivative is as follows: Consider that we plot the graph of the equation $y = f(x)$. We locate the point on the graph corresponding to the assigned value of x and draw the tangent to the curve at that point. The derivative dy/dx is equal to the slope of the tangent. Thus, with reference to Fig. 8.1, if $dy/dx = 0.85$ when $x = a$, the slope of the tangent to the curve at P is 0.85. In general, if θ is the angle between the tangent and the positive side of the x axis, we have $\tan \theta = dy/dx$.

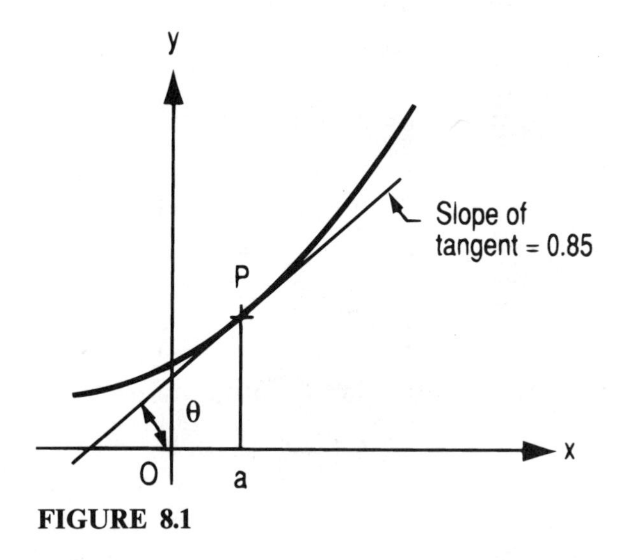

FIGURE 8.1

When a point moves along a curve, the direction of its motion at any instant is along the tangent to the curve at the point it currently occupies. Therefore, dy/dx is the ratio of the instantaneous rate at which y is increasing to the instantaneous rate at which x is increasing.

The process of evaluating a derivative is called *differentiation.* Specifically, when we find dy/dx, we are differentiating the equation for *y with respect to x*. Table A.1 in the appendix presents the derivatives of the basic functions, the derivatives being numbered for convenience.

Now consider that y is a function of v, where v is a function of x. Then

$$\frac{dy}{dx} = \frac{dy}{dv}\frac{dv}{dx} \tag{8.2}$$

This equation can be extended to include any number of intervening variables, and it is known as the *chain rule of derivatives.*

EXAMPLE 8.1 Find dy/dx if $y = \cos^3(3x^2)$.

SOLUTION We first rewrite the equation in this form:

$$y = [\cos(3x^2)]^3$$

We now introduce intervening functions, in this manner:

$$y = u^3 \qquad u = \cos v \qquad v = 3w \qquad w = x^2$$

Applying the chain rule, we have

$$\frac{dy}{dx} = \frac{dy}{du}\frac{du}{dv}\frac{dv}{dw}\frac{dw}{dx}$$

We now refer to Table A.1 and apply Derivatives (6), (8), (4), and again (6), in that order. Then

$$\frac{dy}{dx} = (3u^2)(-\sin v)(3)(2x)$$

Replacing each variable with its expression in terms of x and performing the arithmetic, we finally obtain

$$\frac{dy}{dx} = -18x\cos^2(3x^2)\sin(3x^2)$$

EXAMPLE 8.2 Find dy/dx if

$$y = \frac{5}{(7-2x)^3}$$

SOLUTION We rewrite the equation in this form:

$$y = 5(7-2x)^{-3}$$

Set $$y = 5u^{-3} \qquad u = 7 - 2x$$

Applying the chain rule and Derivatives (4), (6), (2), and (1), in that order, we obtain

$$\frac{dy}{dx} = \frac{dy}{du}\frac{du}{dx} = (-15u^{-4})(-2) = 30u^{-4} = 30(7 - 2x)^{-4}$$

or

$$\frac{dy}{dx} = \frac{30}{(7 - 2x)^4}$$

Where the expression for y is relatively simple, the derivative can be written directly by starting with the outermost function and proceeding to the innermost function.

EXAMPLE 8.3 Find dy/dx if $y = 7x^3 + 5x^2 - 9$.

SOLUTION Applying Derivatives (2), (4), (6), and (1), in that order, we obtain

$$\frac{dy}{dx} = 21x^2 + 10x$$

EXAMPLE 8.4 Find dy/dx if

$$y = \frac{5x^3 - x^2 - 10}{2x + 7}$$

SOLUTION Applying Derivative (5), we obtain

$$\begin{aligned}\frac{dy}{dx} &= \frac{(2x + 7)\dfrac{d(5x^3 - x^2 - 10)}{dx} - (5x^3 - x^2 - 10)\dfrac{d(2x + 7)}{dx}}{(2x + 7)^2}\\ &= \frac{(2x + 7)(15x^2 - 2x) - (5x^3 - x^2 - 10)2}{(2x + 7)^2}\\ &= \frac{20x^3 + 103x^2 - 14x + 20}{(2x + 7)^2}\end{aligned}$$

EXAMPLE 8.5 Find dy/dx if $y = 1/x^5$.

SOLUTION The derivative can be found most readily by rewriting the equation in the form $y = x^{-5}$. Then

$$\frac{dy}{dx} = -5x^{-6} = -\frac{5}{x^6}$$

In general,

$$\frac{d}{dx}\left(\frac{1}{x^n}\right) = -\frac{n}{x^{n+1}} \tag{8.3}$$

8.1.2 Differentiation of Implicit Functions

If the functional relationship between two variables is not stated explicitly, either variable may be considered to be an *implicit function* of the other. We shall illustrate the method of differentiating implicit functions.

EXAMPLE 8.6 Find dy/dx if $x^3y - 2x^2y^2 + 9y^3 = 87$.

SOLUTION For enhanced clarity, we shall enclose the derivative of each term in brackets. Differentiating both sides of the equation with respect to x and applying Derivative (3), we obtain

$$\left[x^3\frac{dy}{dx} + y(3x^2)\right] - 2\left[x^2\left(2y\frac{dy}{dx}\right) + y^2(2x)\right] + \left[27y^2\frac{dy}{dx}\right] = 0$$

Solving for dy/dx, we have

$$\frac{dy}{dx} = \frac{-3x^2y + 4xy^2}{x^3 - 4x^2y + 27y^2}$$

8.1.3 Differentiation with Parametric Equations

In Art. 7.1.5, we defined a parameter and parametric equations. Consider that the variables x and y are both functions of a parameter t. The chain rule of derivatives yields

$$\frac{dy}{dx} = \frac{dy}{dt}\frac{dt}{dx}$$

$$\therefore \quad \frac{dy}{dx} = \frac{dy/dt}{dx/dt} \tag{8.4}$$

EXAMPLE 8.7 A point moves in a plane, and its instantaneous coordinates are $x = 9t^2$, $y = 2t^3 - 5t$, where t denotes elapsed time in seconds. When $t = 3$ s, what is the angle between the direction of motion and the x axis?

SOLUTION As previously stated, the direction of motion is along the tangent to the curve traced by the point. Therefore, we compute dy/dx to find the slope of the tangent, and we apply Eq. (8.4) for that purpose.

$$\frac{dx}{dt} = 18t \qquad \frac{dy}{dt} = 6t^2 - 5$$

$$\frac{dy}{dx} = \frac{6t^2 - 5}{18t}$$

When $t = 3$ s, $dy/dx = 49/54$.

Let θ denote the angle that the tangent to the curve makes with the positive side of the x axis, measured in the counterclockwise direction. Then $\tan\theta = 49/54$, and $\theta = 42.22°$.

8.1.4 Logarithmic Differentiation

Where the expression for y contains x in an exponent, differentiation can be performed by first taking the natural logarithm of both sides of the equation and then differentiating both sides of the logarithmic equation. This process, which is known as *logarithmic differentiation*, requires an application of the laws of logarithms presented in Art. 1.8 and of Derivative (12).

EXAMPLE 8.8 Find dy/dx if $y = x^{\sin 2x}$.

SOLUTION Proceeding in the prescribed manner, we obtain

$$\ln y = (\sin 2x) \ln x$$

$$\frac{1}{y}\frac{dy}{dx} = (\sin 2x)\left(\frac{1}{x}\right) + 2(\cos 2x) \ln x$$

$$\frac{dy}{dx} = y\left[\frac{\sin 2x}{x} + 2(\cos 2x) \ln x\right]$$

8.1.5 Derivatives of Higher Order

We shall now refer to dy/dx as the *first derivative* of y with respect to x. If dy/dx is itself a function of x, it also has a derivative. The derivative of dy/dx is called the *second derivative* of y with respect to x, and it is denoted by d^2y/dx^2, y'', $f''(x)$, or $\ddot{y}$. This process can be continued. In general, the expression obtained by differentiating n times is called the *nth derivative*, and it is denoted by d^ny/dx^n, $y^{(n)}$, or $f^{(n)}(x)$.

EXAMPLE 8.9 Find the fourth derivative of y with respect to x if $y = 3x^6 - 9x^5 + 4x^2 + 8x - 21$.

SOLUTION

$$\frac{dy}{dx} = 18x^5 - 45x^4 + 8x + 8 \qquad \frac{d^2y}{dx^2} = 90x^4 - 180x^3 + 8$$

$$\frac{d^3y}{dx^3} = 360x^3 - 540x^2 \qquad \frac{d^4y}{dx^4} = 1080x^2 - 1080x$$

The second derivative has an extremely important geometric meaning. With reference to Fig. 8.1, dy/dx equals the slope of the tangent to the curve at P. Consider that we move along the curve in such manner that x increases at a uniform rate. Since d^2y/dx^2 equals the rate at which dy/dx is increasing, it also equals the rate at which the slope of the tangent to the curve is increasing during this movement.

8.1.6 Hyperbolic Functions

We shall now define a class of functions that plays an important role in calculus. Because these functions arise in the study of the geometry of the hyperbola, they are called *hyperbolic functions.* The notation for a given function is formed by adding the letter h to the analogous circular function defined in Art. 6.2.3. For example, the notation sinh denotes *hyperbolic sine.* The hyperbolic functions involve the quantity e, which is defined in Art. 1.20. The three basic functions are as follows:

$$\sinh u = \frac{e^u - e^{-u}}{2}$$

$$\cosh u = \frac{e^u + e^{-u}}{2}$$

$$\tanh u = \frac{\sinh u}{\cosh u} = \frac{e^u - e^{-u}}{e^u + e^{-u}}$$

8.1.7 Differentials

In Fig. 8.2, P is a point on curve C, Q is a neighboring point on C and to the right of P, and T is the tangent to the curve at P. Lines AP and BQ are parallel to the y axis, and line PD is parallel to the x axis. Let Δx and Δy denote the increment of x and of y, respectively, as we move from P to Q.

We shall now replace the notation Δx with dx, and we shall call this the *differential of x.* If y were increasing along line T rather than curve C, the increment of y corresponding to dx would be DR. We label this quantity dy, and we call it the *differential of y.* Since the derivative of y with respect to x equals the slope of the tangent to the curve, the notation dy/dx can refer either to the first derivative or to the ratio of the two differentials, the two quantities being equal. Letting $f'(x)$ denote the first derivative, we have

$$dy = f'(x)\, dx \tag{8.5}$$

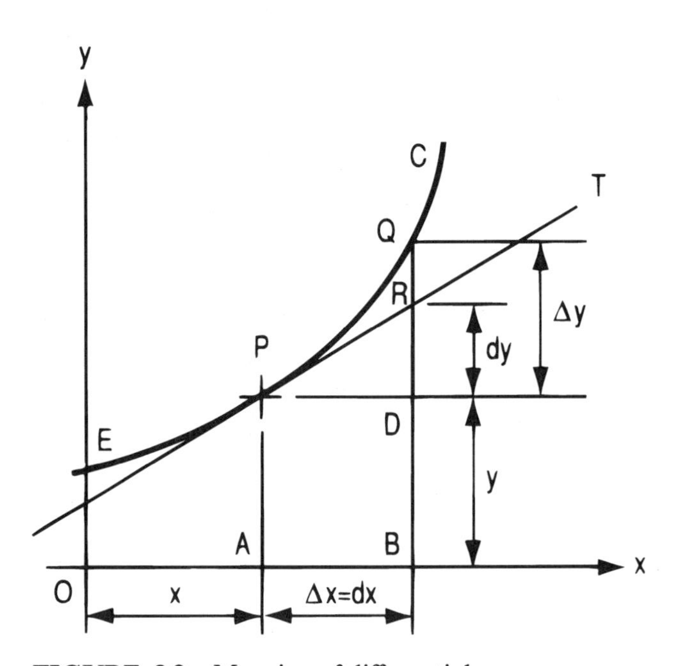

FIGURE 8.2 Meaning of differentials.

If dx is extremely small, dy is a close approximation to the true increase in y, which is Δy.

EXAMPLE 8.10 If $y = 3x^4$, what is dy when $x = 1.5$ and $dx = 0.003$? What is the amount (to five significant figures) by which y increases when x increases from 1.5 to 1.503?

SOLUTION

$$f'(x) = 12x^3 \qquad dy = 12(1.5)^3(0.003) = 0.1215$$

$$\Delta y = 3[(1.503)^4 - (1.5)^4] = 0.12186$$

8.1.8 Derivative of Arc Length and Area

With reference to Fig. 8.2, consider that a line starts at the y axis and then advances to the right under these conditions: It remains parallel to the y axis, and it terminates on curve C. As the line advances from AP to BQ, its terminus generates the arc PQ, and the line generates the area bounded by line AP, line BQ, the x axis, and curve C.

Let s denote the length of the arc EP. Then Δs is the length of the arc PQ. The *chord PQ* has this relationship:

$$(PQ)^2 = (\Delta x)^2 + (\Delta y)^2$$

As Δx approaches zero, Δs approaches chord PQ, and it follows that

$$\frac{ds}{dx} = \sqrt{1 + \left(\frac{dy}{dx}\right)^2} \tag{8.6}$$

Now let A denote the area $OAPE$ generated by the moving line. Then ΔA is the area $ABQP$. This consists of the area $ABDP$, which equals $(AP)\ \Delta x$, and the area PDQ. As Δx approaches zero, the latter area approaches zero, and it follows that

$$\frac{dA}{dx} = y \tag{8.7}$$

EXAMPLE 8.11 Formulate the expression for ds/dx and for dA/dx for the curve having the equation $y = 1/x^2$.

SOLUTION The curve is plotted in Fig. 8.3. It consists of two branches that are symmetrical about the y axis, and each branch is asymptotic to the x and y axes. Applying Eq. (8.3), we have

$$\frac{dy}{dx} = -\frac{2}{x^3}$$

Then

$$\frac{ds}{dx} = \sqrt{1 + \frac{4}{x^6}} \qquad \frac{dA}{dx} = y = \frac{1}{x^2}$$

Consider that we start with $x = 0$ and then let x increase. From the graph, we

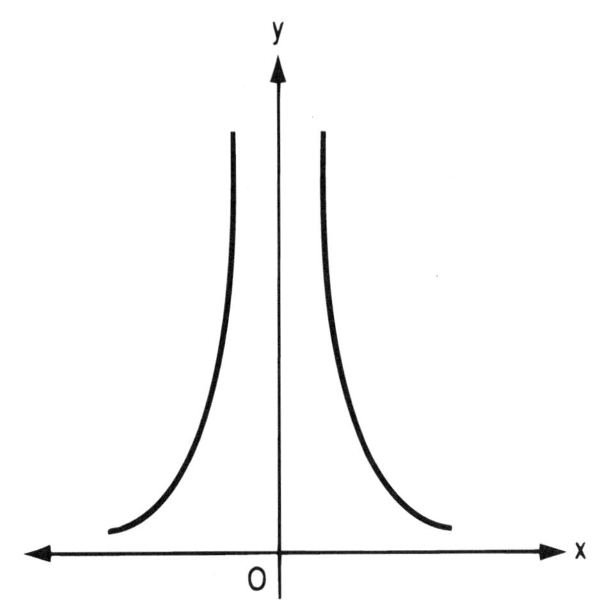

FIGURE 8.3 Graph of $y = 1/x^2$.

glean these facts: When x is very small, both s and A increase with extreme rapidity. As x becomes infinitely large, the curve approaches parallelism with the x axis; therefore, ds/dx approaches 1 as a limit. Also, as x becomes infinitely large, A increases negligibly. The equations for ds/dx and for dA/dx are consonant with these observed facts.

8.2 APPLICATIONS OF DERIVATIVES

8.2.1 Properties of a Curve

The first and second derivatives of a function enable us to identify the properties of the graph of the function. We shall refer to Fig. 8.4 for illustrative purposes.

We first present the following definitions: A curve is *concave upward* if a chord connecting two points on the curve lies above the curve, and it is *concave downward* if the chord lies below the curve. Thus, the curve in Fig. 8.4 is concave upward to the left of Q and concave downward to the right of Q. The point Q at which the concavity changes is a *point of inflection.* Points P_1 and P_2 are the *turning points* of the curve. Point P_2, where y has a local maximum value, is a *summit*; point P_1, where y has a local minimum value, is a *nadir*.

To identify the properties of the curve, we apply these principles: dy/dx equals the slope of the tangent to the curve, and d^2y/dx^2 equals the rate at which this slope is changing. To the left of Q, the slope of the tangent changes from negative to positive, and therefore the slope is *increasing*; to the right of Q, the slope of the tangent changes from positive to negative, and therefore the slope is *decreasing*. The following principles emerge:

1. In a region where the curve is concave upward, d^2y/dx^2 is positive; in a region where the curve is concave downward, d^2y/dx^2 is negative.
2. At a point of inflection, $d^2y/dx^2 = 0$.

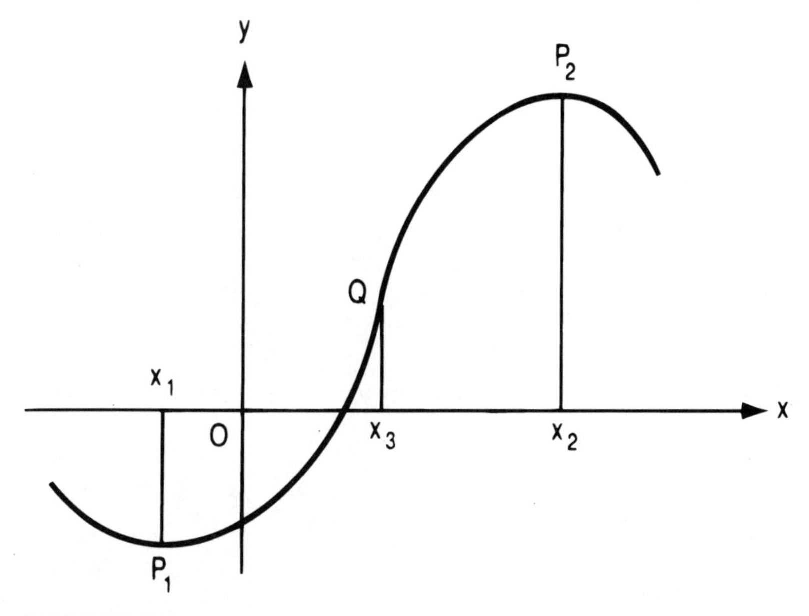

FIGURE 8.4

3. At a summit, $dy/dx = 0$ and d^2y/dx^2 is negative.
4. At a nadir, $dy/dx = 0$ and d^2y/dx^2 is positive.

EXAMPLE 8.12 A curve has the equation $y = 2x^3 - 6x^2 - 90x$. Locate its significant points.

SOLUTION The curve appears in Fig. 8.5.

$$\frac{dy}{dx} = 6x^2 - 12x - 90 = 6(x^2 - 2x - 15) = 6(x + 3)(x - 5)$$

$$\frac{d^2y}{dx^2} = 12x - 12$$

Setting $dy/dx = 0$, we obtain $x = -3$ and $x = 5$. When $x = -3$, $d^2y/dx^2 = -48$;

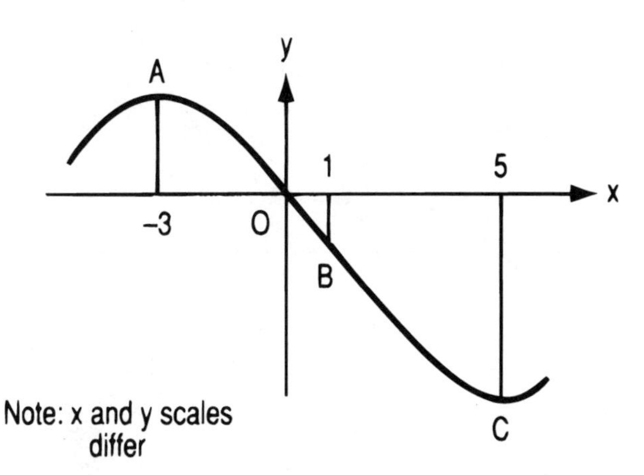

FIGURE 8.5 Graph of $y = 2x^3 - 6x^2 - 90x$.

when $x = 5$, $d^2y/dx^2 = 48$. Finally, setting $d^2y/dx^2 = 0$, we obtain $x = 1$. We thus arrive at the following conclusions:

The curve has a summit at A, where $x = -3$, and it has a nadir at C, where $x = 5$. It has a point of inflection at B, where $x = 1$.

The general cubic curve is skew-symmetrical about a line through its point of inflection. Specifically, in Fig. 8.5, consider that the curve is transformed in this manner: Pass a line through B parallel to the y axis; then revolve the part of the curve to the right of this line about the x axis through an angle of 90°. The two parts of the curve are now symmetrical about this line.

EXAMPLE 8.13 A curve has the equation $y = x^3 - 12x^2 + 75x$. Locate its significant points.

SOLUTION The curve appears in Fig. 8.6*a*.

$$\frac{dy}{dx} = 3x^2 - 24x + 75 \qquad \frac{d^2y}{dx^2} = 6x - 24$$

When we set $dy/dx = 0$, we obtain roots that are complex numbers. Therefore, this curve has no turning points. When we set $d^2y/dx^2 = 0$, we obtain $x = 4$. Therefore, the curve has a point of inflection at A.

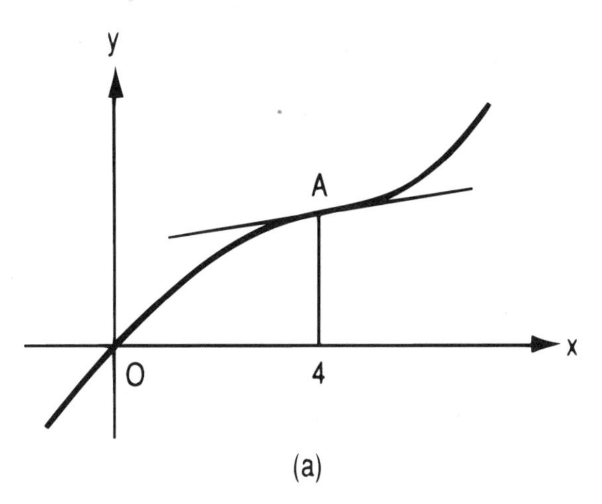

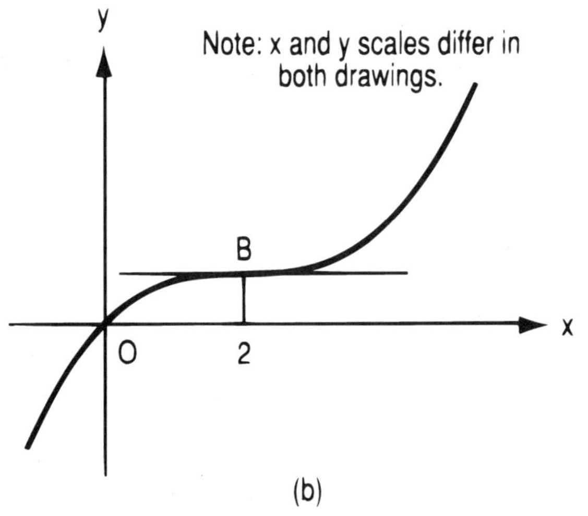

FIGURE 8.6 (*a*) Graph of $y = x^3 - 12x^2 + 75x$; (*b*) graph of $y = x^3 - 6x^2 + 12x$.

EXAMPLE 8.14 A curve has the equation $y = x^3 - 6x^2 + 12x$. Locate its significant points.

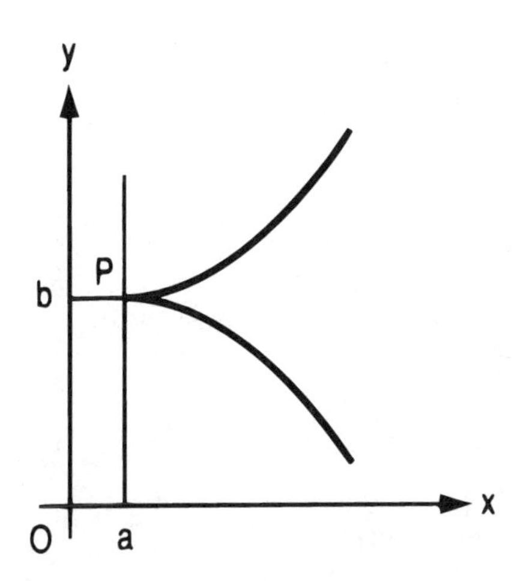

FIGURE 8.7 Graph of $y = (x - a)^{3/2} + b$.

SOLUTION The curve appears in Fig. 8.6*b*. When we set $dy/dx = 0$ and $d^2y/dx^2 = 0$, we obtain the same result: $x = 2$. Therefore, the curve has no turning points, and it has a point of inflection at *B*, where the tangent to the curve is horizontal.

Another special type of point that a curve may have is a *cusp point*, which is illustrated by point *P* in Fig. 8.7. At a cusp point, the curve is divided precipitously into two branches, and consequently d^2y/dx^2 is infinite in absolute value. However, both x and y have finite values.

EXAMPLE 8.15 A curve has the equation $y = (x - a)^{3/2} + b$, where a and b are constants. Locate its cusp point.

SOLUTION The equation may be rewritten in this form:

$$y = \sqrt{(x - a)^3} + b$$

Since y is restricted to real values, x is subject to the restriction $x \geq a$. The curve is shown in Fig. 8.7.

$$\frac{dy}{dx} = \frac{3}{2}(x - a)^{1/2} \qquad \frac{d^2y}{dx^2} = \frac{3}{4}(x - a)^{-1/2} = \frac{3}{4\sqrt{x - a}}$$

When $x = a$, $y = b$, but d^2y/dx^2 is infinite. Therefore, the curve has a cusp at *P*, as shown.

8.2.2 Maxima and Minima

Again let y denote a function of x. Article 8.2.1 reveals that the derivatives of y enable us to identify the values of x at which y has a maximum or minimum value. The procedure consists of setting up the expression for y in terms of x, obtaining the expression for dy/dx, setting $dy/dx = 0$, and then solving for x. Whether the corresponding value of y is maximum or minimum is usually self-evident. However, if such is not the case, the issue can be resolved by computing d^2y/dx^2 for the computed value of x. The value of y is maximum if d^2y/dx^2 is negative, and it is minimum if d^2y/dx^2 is positive.

EXAMPLE 8.16 Let b denote the width and d the depth of a beam of rectangular cross section. The strength of the beam is directly proportional to bd^2. Establish the dimensions of the strongest timber beam that can be cut from a cylindrical log having a diameter D.

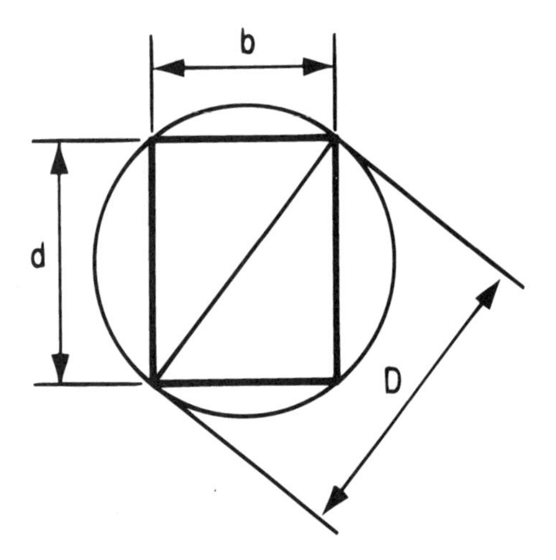

FIGURE 8.8 Dimensions of a beam cut from a cylindrical log.

SOLUTION Refer to Fig. 8.8. The variables have the relationship

$$b^2 + d^2 = D^2$$

Let S denote the strength of the beam and k denote the constant of proportionality. We shall express S as a function of b. Then

$$S = kbd^2 = kb(D^2 - b^2) = k(D^2b - b^3)$$

$$\frac{dS}{db} = k(D^2 - 3b^2)$$

Setting $dS/db = 0$ and solving for b, we obtain the following as the dimensions of the strongest beam:

$$b = \sqrt{\frac{1}{3}}\,D \qquad d = \sqrt{\frac{2}{3}}\,D$$

EXAMPLE 8.17 With reference to Example 8.16, the *stiffness* of the beam is directly proportional to bd^3. Establish the dimensions of the stiffest beam that can be cut.

SOLUTION Let T denote the stiffness and k denote the constant of proportionality. Then

$$T = kbd^3 = kb(D^2 - b^2)^{3/2}$$

Applying Derivative (3), we obtain

$$\begin{aligned}\frac{1}{k}\frac{dT}{db} &= b\left(\frac{3}{2}\right)(D^2 - b^2)^{1/2}(-2b) + (D^2 - b^2)^{3/2} \\ &= -3b^2(D^2 - b^2)^{1/2} + (D^2 - b^2)^{3/2} \\ &= (D^2 - b^2)^{1/2}(-3b^2 + D^2 - b^2)\end{aligned}$$

Setting $dT/db = 0$, we obtain two results: $b = D$, which lacks physical significance, and $b = D/2$. Therefore, the required dimensions are

$$b = \frac{D}{2} \qquad d = \frac{D}{2}\sqrt{3}$$

EXAMPLE 8.18 In Fig. 8.9, a light source is placed at A and at B, where $AB = 6$ m. The source at A is 4 times stronger than that at B. An object is to be placed on line AB. If the intensity of illumination at a given point is inversely pro-

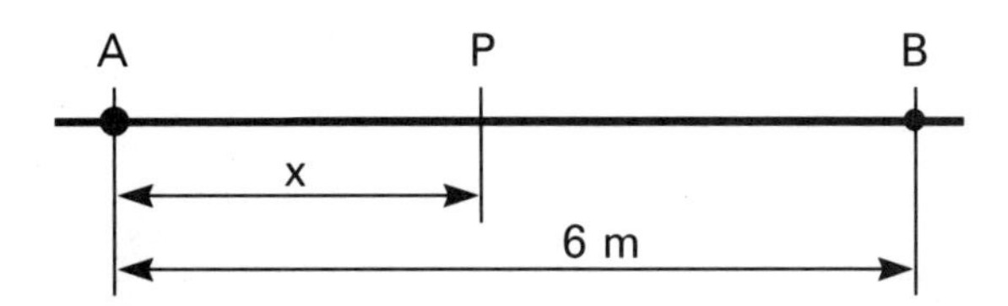

FIGURE 8.9 Determining position of an object to minimize its illumination.

portional to the square of the distance from the source to the point, where shall the object be placed to receive minimum illumination?

SOLUTION Let P denote the point at which the object is placed. Let $AP = x$; then $PB = 6 - x$. Now let I denote the intensity of illumination at P and k denote the constant of proportionality. Then

$$I = k\left[\frac{4}{x^2} + \frac{1}{(6-x)^2}\right]$$

or

$$I = k[4x^{-2} + (6-x)^{-2}]$$

$$\frac{dI}{dx} = k[4(-2)x^{-3} + (-2)(6-x)^{-3}(-1)]$$

$$= k\left[-\frac{8}{x^3} + \frac{2}{(6-x)^3}\right]$$

Setting $dI/dx = 0$, we obtain

$$\left(\frac{x}{6-x}\right)^3 = 4 \qquad \frac{x}{6-x} = 4^{1/3} = 1.5874$$

Solving, we obtain $x = 3.6811$ m.

In general, if $AB = a$ and r is the ratio of the strength of the source at A to that at B, we have

$$x = a\,\frac{r^{1/3}}{r^{1/3} + 1}$$

If $r = 1$, $x = a/2$. As r increases, x increases. Theoretically, if r increases beyond bound, x approaches a as a limit.

EXAMPLE 8.19 In Fig. 8.10, line AB is to be drawn through point $P(a, b)$. Lines CP and DP are parallel to the x and y axes, respectively. Find the lengths of OA and OB if the area OAB is to be minimum. At this condition, what is the value of angle θ? What is the ratio of area CPB to area DAP?

SOLUTION Let A denote the area OAB. Consider that AB is placed in an arbitrary position and then rotated about P. As θ approaches either 0 or 90°, A becomes infinitely large. Our problem is to find the specific position of AB at which A is minimum. Set $DA = u$ and $CB = v$.

$$OD = CP = a \qquad OC = DP = b$$

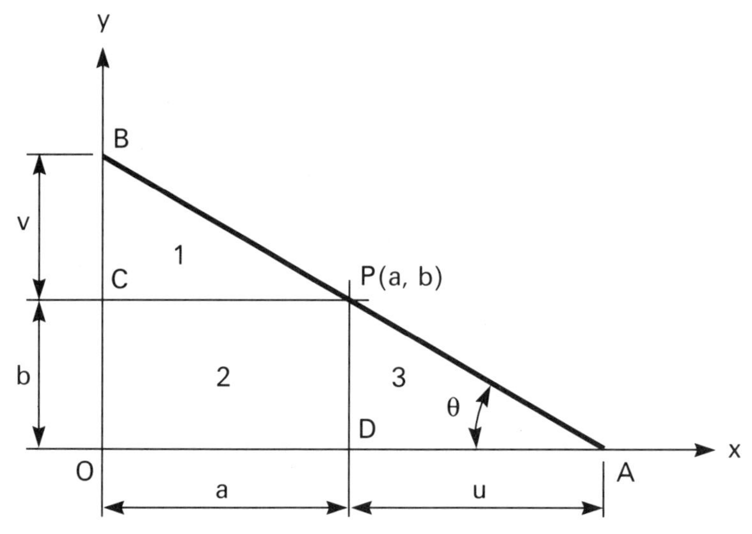

FIGURE 8.10 Determining position of line AB to minimize area OAB.

Triangles CPB and DAP are similar because their corresponding angles are equal. Therefore,

$$\frac{CB}{CP} = \frac{DP}{DA} \qquad \text{or} \qquad \frac{v}{a} = \frac{b}{u}$$

Then $v = ab/u$.

For simplicity, divide the total area into the three parts that are labeled in the diagram. The areas of the individual parts are as follows:

$$A_1 = \frac{av}{2} = \frac{a}{2}\frac{ab}{u} = \frac{a^2b}{2u}$$

$$A_2 = ab \qquad A_3 = \frac{bu}{2}$$

and the total area is

$$A = \frac{a^2b}{2u} + ab + \frac{bu}{2}$$

Then

$$\frac{dA}{du} = -\frac{a^2b}{2u^2} + \frac{b}{2}$$

Setting $dA/du = 0$, we obtain $u = a$, and it follows that $v = b$. Thus, to minimize the area OAB, we set $OA = 2a$, and OB becomes $2b$. At this condition, P is the midpoint of AB, and $AP = OP$. Triangles DAP and CPB are congruent; therefore, their areas are equal. Angle $\theta = \arctan b/a$.

An alternative method of solution consists of expressing A as a function of θ. On setting $dA/d\theta = 0$, we obtain $\tan \theta = b/a$, and it follows that $DA = a$.

The conclusion reached in Example 8.19 regarding the equality of the areas DAP and CPB is a general one. Let x denote the independent variable, and assume that a quantity S is the sum of three parts: one that varies directly with x, one that varies inversely with x, and one that is independent of x. Then S is minimum when the first and second parts are equal. This relationship arises in engineering economics, where it is known as *Kelvin's law*.

EXAMPLE 8.20 An equilateral hyperbola has the equation $x^2 - y^2 = 144$. Find the minimum distance from the point $P(0, 4)$ to the curve.

SOLUTION Let $Q(x, y)$ denote an arbitrary point on the curve, and let $u = PQ$. By Eq. (7.8),

$$u^2 = x^2 + (y - 4)^2 = (144 + y^2) + (y^2 - 8y + 16) = 2y^2 - 8y + 160$$

$$\frac{du^2}{dy} = 4y - 8$$

Setting the derivative equal to zero, we obtain $y = 2$; then $x = \pm\sqrt{148}$. These are the coordinates of the point on a branch of the curve at which the distance is minimum. Then

$$u = \sqrt{148 + (-2)^2} = \sqrt{152} = 12.3288$$

Refer to Fig. 7.14. In the present instance. the vertex of the hyperbola lies at a distance of 12 from the origin, and the distance from P to the vertex is

$$\sqrt{4^2 + 12^2} = \sqrt{160} = 12.6491$$

The minimum distance from P to the curve is slightly less than this amount.

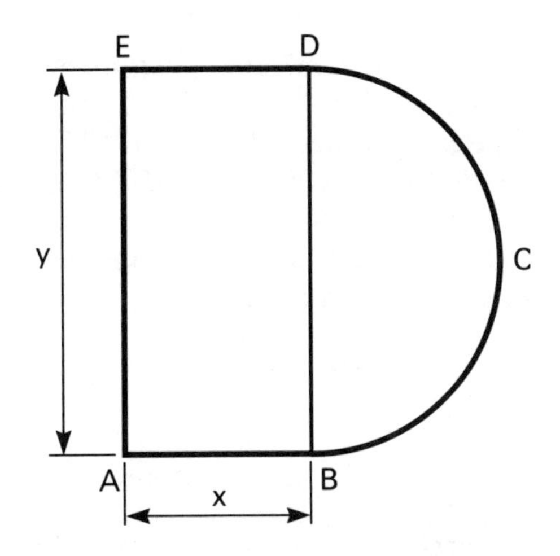

FIGURE 8.11 Plot to be enclosed with fencing.

EXAMPLE 8.21 With reference to Fig. 8.11, a plot of land consisting of a rectangle $ABDE$ and a semicircle BCD is to be enclosed with fencing, and the amount of fencing available is 200 m. Determine the values of x and y that will maximize the enclosed area, and compute the maximum area.

SOLUTION The perimeter and area of a *complete* circle are $2\pi r$ and πr^2, respectively, where r denotes the radius. Computing the amount of fencing required and equating it to 200, we obtain

$$2x + y + \pi r = 2x + y + \pi\left(\frac{y}{2}\right) = 200$$

Let A denote the area enclosed by the fencing. Then

$$A = xy + \frac{\pi r^2}{2} = xy + \frac{\pi(y/2)^2}{2} = xy + \frac{\pi y^2}{8}$$

To simplify the calculations, we shall express A as a function of y. The first equation yields

$$x = 100 - \left(\frac{2 + \pi}{4}\right)y$$

Substituting this expression in the expression for A and simplifying, we obtain

$$A = 100y - \left(\frac{4 + \pi}{8}\right)y^2$$

Then
$$\frac{dA}{dy} = 100 - \left(\frac{4 + \pi}{4}\right)y$$

Equating this derivative to 0, we obtain $y = 56.01$ m, and it follows that $x = 28.00$ m. The maximum area is 2800 m^2. In general, in a problem of this type, $y = 2x$ when the area is maximum.

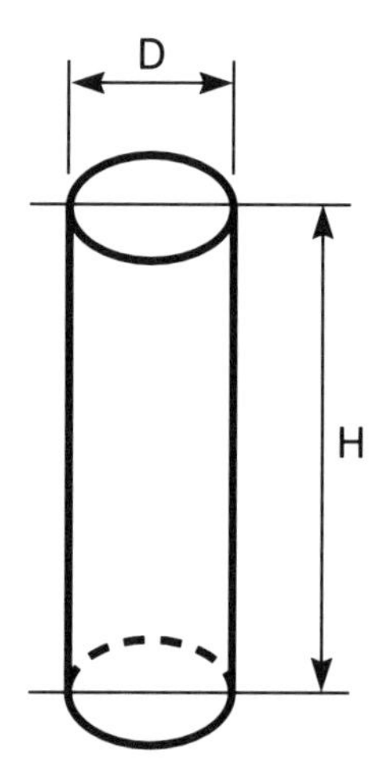

FIGURE 8.12 Dimensions of a cylindrical tank.

EXAMPLE 8.22 The cylindrical tank in Fig. 8.12 is to have a volume of 6 m^3, and it is to be completely covered. The covering material that is applied to the top and bottom is 3 times as costly as that which is applied to the side. Determine the values of H and D that will minimize the cost of covering the tank.

SOLUTION The total area of the top and bottom is $2(\pi D^2/4) = \pi D^2/2$, and the surface area of the side is πDH. The volume V of the tank is $(\pi D^2/4)H$. Let C denote the cost of covering the tank and k denote the constant of proportionality. We shall express C in terms of D.

$$V = \frac{\pi D^2 H}{4} = 6$$

Then
$$H = \frac{24}{\pi D^2} \quad \text{and} \quad DH = \frac{24}{\pi D}$$

$$C = k\left[3\left(\frac{\pi D^2}{2}\right) + \pi DH\right] = k\left(1.5\pi D^2 + \frac{24}{D}\right)$$

$$\frac{dC}{dD} = k\left(3\pi D - \frac{24}{D^2}\right)$$

Equating this derivative to 0, we obtain $D = (8/\pi)^{1/3} = 1.3656$ m, and it follows that $H = 4.0965$ m.

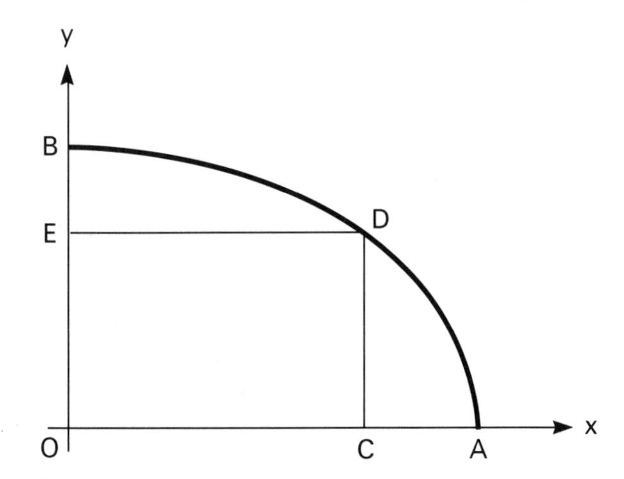

FIGURE 8.13 Inscribed rectangle.

It is interesting to observe that $H/D = 3$, and this is the ratio of the cost of the top-and-bottom material to that of the side material. It can be demonstrated that this equality is a general one.

EXAMPLE 8.23 The quarter-ellipse in Fig. 8.13 has the equation $x^2 + Ay^2 = B$, where A and B are constants. A rectangle $OCDE$ is to be constructed, with its vertex D on the curve. Determine the dimensions of the rectangle that will maximize its area.

SOLUTION Let $OC = x$ and $OE = y$; let A denote the area of the rectangle.

$$x = (B - Ay^2)^{1/2} \qquad A = xy = y(B - Ay^2)^{1/2}$$

$$\frac{dA}{dy} = y\left[\frac{1}{2}(B - Ay^2)^{-1/2}(-2Ay)\right] + (B - Ay^2)^{1/2}$$

$$= -\frac{Ay^2}{(B - Ay^2)^{1/2}} + (B - Ay^2)^{1/2}$$

Equating this derivative to 0, we obtain $y = \sqrt{B/2A}$, and it follows that $x = \sqrt{B/2}$.

8.2.3 Related Rates of Change

Assume the following: The quantities x and y are related, they both vary with time, we are given the rate at which x is changing, and we wish to determine the rate at which y is changing. Let t denote elapsed time. Equation (8.2) becomes

$$\frac{dy}{dt} = \frac{dy}{dx}\frac{dx}{dt} \tag{8.8}$$

Thus, to find dy/dt, it is necessary to express y in terms of x and to differentiate the resulting expression.

EXAMPLE 8.24 An ideal gas undergoes a polytropic process having the equation

$$PV^{1.36} = C \tag{a}$$

where P and V denote absolute pressure and volume, respectively, and C is a constant. At a given instant, $P = 0.35$ Pa, $V = 0.13\ \text{m}^3$, and the volume is increasing at the rate of $0.006\ \text{m}^3/\text{s}$. Find the rate at which the pressure is changing.

SOLUTION We rewrite Eq. (*a*) as $P = CV^{-1.36}$ Then

$$\frac{dP}{dV} = -1.36CV^{-2.36} = -1.36(PV^{1.36})V^{-2.36} = -\frac{1.36P}{V}$$

$$\frac{dP}{dt} = \frac{dP}{dV}\frac{dV}{dt} = -\frac{1.36(0.35)}{0.13}(0.006) = -0.02197\ \text{Pa/s}$$

EXAMPLE 8.25 A ladder 6 m long rests against the wall of a building. At a given instant, the base of the ladder is 1.5 m from the wall, and it is slipping outward at the rate of 0.9 m/s. How fast is the upper end of the ladder descending? Devise an approximate test of the result.

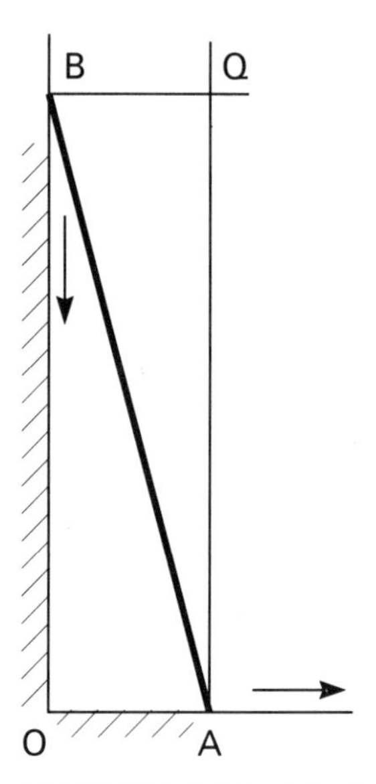

FIGURE 8.14 Motion of a ladder.

SOLUTION Refer to Fig. 8.14. Let $x = OA$ and $y = OB$. Then

$$x^2 + y^2 = 6^2 \qquad \frac{dx}{dt} = 0.9\ \text{m/s}$$

$$y = (36 - x^2)^{1/2}$$

$$\frac{dy}{dx} = \frac{1}{2}(36 - x^2)^{-1/2}(-2x)$$

$$= -\frac{x}{\sqrt{36 - x^2}}$$

$$= -\frac{1.5}{\sqrt{36 - 1.5^2}} = -0.2582$$

$$\frac{dy}{dt} = \frac{dy}{dx}\frac{dx}{dt} = (-0.2582)(0.9)$$

$$= -0.2324\ \text{m/s}$$

Thus, the upper end of the ladder is descending at the rate of 0.2324 m/s.

This result can be tested approximately by assuming that the rates of change of x and y remain constant for a very brief interval of time, and we select 1/1000 s for this purpose. The displacement of A during this interval would be 0.0009 m. The initial value of x is 1.5 m, and its final value would be 1.5009 m. Let y_i and y_f denote the initial and final values of y, respectively. Then

$$y_i = \sqrt{6^2 - 1.5^2} \qquad y_f = \sqrt{6^2 - 1.5009^2}$$

Taking the difference between the two y values and multiplying the result by 1000, we obtain 0.2325 m to four significant figures. This value is extremely close to the calculated rate of change.

Example 8.25 can be solved in a simpler manner by applying the principles of mechanics. In Fig. 8.14, draw a vertical line through A and a horizontal line through B, and call the intersection point Q. Since A is moving horizontally and B is moving vertically, Q is the instantaneous center of rotation of AB. Let v_A and v_B denote the instantaneous velocity of A and of B, respectively. In absolute value,

$$\frac{v_B}{v_A} = \frac{BQ}{AQ} \qquad (b)$$

$$v_A = 0.9 \text{ m/s} \qquad BQ = 1.5 \text{ m}$$

$$AQ = \sqrt{6^2 - 1.5^2} = 5.8095 \text{ m}$$

Substituting in Eq. (b), we obtain $v_B = 0.2324$ m/s.

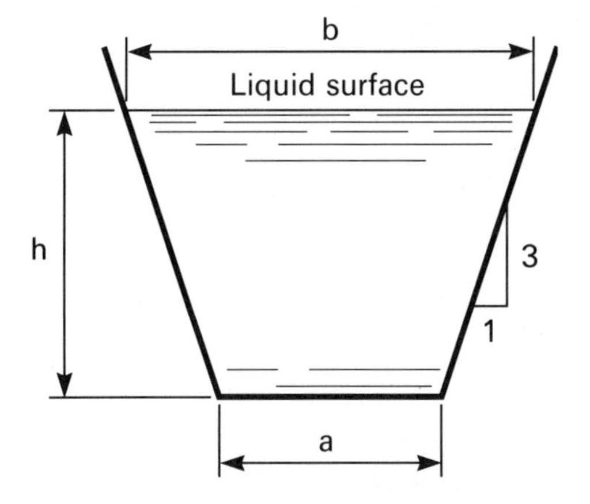

FIGURE 8.15 Cross section of liquid tank.

EXAMPLE 8.26 A tank is in a horizontal position, and its length is L. Its vertical cross section is an isosceles trapezoid having the dimensions shown in Fig. 8.15. Establish the relationship between the rate at which the liquid surface is rising and the rate at which liquid is entering the tank.

SOLUTION With reference to the liquid that is in the tank at a given instant, let h, A, and V denote its height, its cross-sectional area, and its volume, respectively. The width b of the liquid surface is

$$b = a + \frac{2}{3}h \qquad \text{and} \qquad A = \left(\frac{a+b}{2}\right)h = ah + \frac{h^2}{3}$$

$$V = AL = L\left(ah + \frac{h^2}{3}\right) = \frac{L}{3}(3ah + h^2)$$

$$\frac{dV}{dh} = \frac{L}{3}(3a + 2h) \qquad \text{and} \qquad dh = \frac{3}{L(3s + 2h)}\,dV$$

The rate at which liquid is entering the tank is dV/dt. From the last result, we have

$$\frac{dh}{dt} = \frac{3}{L(3a + 2h)}\frac{dV}{dt}$$

8.2.4 Evaluation of Indeterminate Forms

Let $f(x)$ and $g(x)$ denote functions of x. Then $f'(x)$ and $g'(x)$ denote their respective first derivatives. Consider that we wish to evaluate $f(x)/g(x)$ when $x = a$. If direct substitution of a for x yields the indeterminate form 0/0 or ∞/∞, this obstacle can be circumvented by applying the following equation:

$$\lim_{x \to a} \frac{f(x)}{g(x)} = \lim_{x \to a} \frac{f'(x)}{g'(x)} \tag{8.9}$$

This relationship is referred to as *l'Hôpital's rule*. If the expression at the right also yields an indeterminate form, we take the ratio of the second derivatives, continuing the process until a determinate form is attained. Let $y = f(x)/g(x)$. For brevity, we shall refer to the limiting value of y as x approaches a simply as the value of y when $x = a$.

EXAMPLE 8.27 Evaluate the following function when $x = 0$:

$$y = \frac{1 - \cos x}{x^2}$$

SOLUTION Direct substitution of 0 for x yields 0/0. Therefore, we apply Eq. (8.9) to obtain

$$y = \frac{\sin x}{2x}$$

This form also yields 0/0 when $x = 0$. Therefore, we apply Eq. (8.9) again to obtain

$$y = \frac{\cos x}{2}$$

When $x = 0$, $y = 1/2 = 0.5$. As a test, we may set $x = 0.04$ rad, and we find that $y = 0.49993$, which is extremely close to 0.5.

Other indeterminate forms can be manipulated to make them amenable to l'Hôpital's rule by transforming them to fractions. We shall illustrate the procedure.

EXAMPLE 8.28 Evaluate the function $y = x \ln x$ when $x = 0$.

SOLUTION Direct substitution yields the indeterminate form $0(-\infty)$. However, we may rewrite the given expression in this form:

$$y = \frac{\ln x}{1/x}$$

Applying l'Hôpital's rule, we obtain

$$y = \frac{1/x}{-1/x^2} = -x$$

When $x = 0$, $y = 0$. As a test, we may set $x = 0.0001$. Then $y = -0.00092$, and our result appears to be confirmed.

EXAMPLE 8.29 Evaluate the function $y = \sec x - \tan x$ when $x = \pi/2$ rad.

SOLUTION Direct substitution yields the indeterminate form $\infty - \infty$. However, we may express y in the following alternative manner:

$$y = \frac{1}{\cos x} - \frac{\sin x}{\cos x} = \frac{1 - \sin x}{\cos x}$$

Applying l'Hôpital's rule, we obtain

$$y = \frac{-\cos x}{-\sin x} = \frac{1}{\tan x}$$

When $x = \pi/2$, $y = 0$. As a test, we find that when x is very close to $\pi/2$, y is very close to 0.

EXAMPLE 8.30 Evaluate the function $y = x^x$ when $x = 0$.

SOLUTION Direct substitution yields the indeterminate form 0^0. Taking the natural logarithm of both sides, we obtain

$$\ln y = x \ln x$$

In Example 8.28, we found that the expression at the right is 0 when $x = 0$. Therefore, $\ln y = 0$, and $y = 1$. As a test, we may set $x = 0.001$, and we find that $y = 0.9931$, which is extremely close to 1.

8.2.5 Curvature

Consider that a point moves along a curve from right to left at constant speed. At every instant, the *direction* of its motion is along the tangent to the curve at the point it currently occupies, but the inclination of the tangent is continuously changing. The rate at which the inclination is changing as the point moves is termed the *curvature* of the curve.

Let K denote the curvature of a curve at a given point. In Fig. 8.16, T is the tangent to the curve S at P, θ is the angle that T makes with the positive side of the x axis, and s is the length of arc AP. The curvature is defined as $d\theta/ds$; thus, the unit of curvature is radians per unit of length. Let y' and y'' denote, respectively, the first and second derivative of y with respect to x. Then

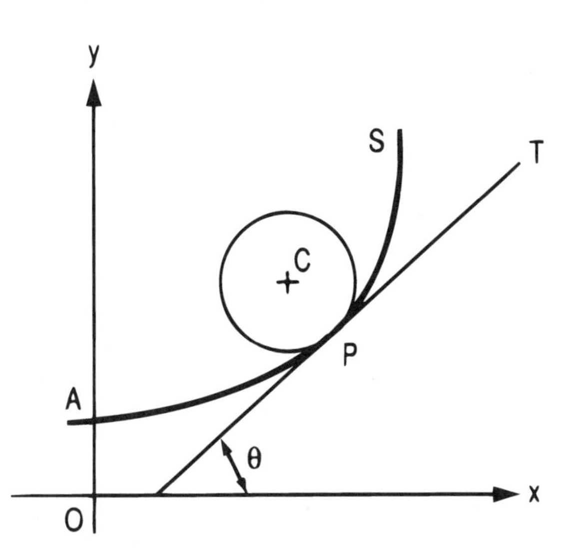

FIGURE 8.16 Curvature, circle of curvature, and center of curvature.

$$K = \frac{y''}{[1 + (y')^2]^{3/2}} \tag{8.10}$$

The algebraic sign of K is identical with that of y'', but usually it is only the absolute value of K that is significant.

With reference to a straight line, y' is constant, and therefore $y'' = 0$. Thus, Eq. (8.10) yields $K = 0$ for a straight line, as it must.

EXAMPLE 8.31 What is the curvature of the parabola $y = x^2$ at the origin? What limit does the curvature approach as x becomes infinitely large in absolute value?

SOLUTION

$$y' = 2x \qquad y'' = 2$$

$$K = \frac{2}{(1 + 4x^2)^{3/2}}$$

When $x = 0$, $K = 2$, and that is the maximum value of K. As x becomes infinitely large in absolute value, K approaches the limiting value of 0. This value signifies that the curve approaches linearity as it recedes from the y axis.

EXAMPLE 8.32 Complete the curvature of the curve $y = \sin x$ when $x = 0$ and when $x = \pi/2$.

SOLUTION

$$y' = \cos x \qquad y'' = -\sin x$$

When $x = 0$, $y' = 1$ and $y'' = 0$. When $x = \pi/2$, $y' = 0$ and $y'' = -1$. Equation (8.10) yields these values: When $x = 0$, $K = 0$; when $x = \pi/2$, $K = -1$. Thus, in absolute value, the curvature of a sine curve ranges from 0 to 1. In the vicinity of $x = 0$, the curve approximates a straight line.

Let R denote the radius of a circle. By simple geometry, we find that the curvature of a circle is uniform along its length (as is intuitively apparent) and that it equals $1/R$ in absolute value. By extension, we define the *radius of curvature* R of a curve in general in this manner:

$$R = \frac{1}{K} \tag{8.11}$$

In Fig. 8.16, we draw a circle on the concave side of the curve and tangent to T (and therefore curve S) at P, the radius of the circle being equal to the radius of curvature of S at P. This circle is the *circle of curvature* of S at P, and its center C is the *center of curvature* of S at P. In general, the centers of curvature of S form a smooth curve that is called the *evolute* of S.

8.3 PARTIAL DIFFERENTIATION

8.3.1 Definition and Notation

Thus far, we have investigated situations where a given variable is a function of a single variable. We now extend the scope of our investigation to situations where a given variable is a function of two independent variables, and the principles we shall develop are applicable to situations where the number of independent variables exceeds two.

Let x and y denote quantities that vary independently of each other, and let z denote a function of x and y. Expressed symbolically, $z = f(x,y)$. The graph of this equation is a surface such that each point on the surface has coordinates that satisfy the given equation.

Although in general x and y vary simultaneously, it is helpful to isolate the effects that their variations have on z. Consider that we hold y constant and allow x to vary. The rate of change of z with respect to x is termed the *partial derivative* of z with respect to x, and it is denoted by $\partial z/\partial x$. Now consider that we hold x constant and allow y to vary. The rate of change of z with respect to y is termed the partial derivative of z with respect to y, and it is denoted by $\partial z/\partial y$.

The geometric interpretation of partial derivatives is as follows: Let S denote the surface having the equation $z = f(x,y)$. Figure 8.17 shows the part of S that lies in the first octant. Let $P(x_1,y_1,z_1)$ denote a point on this surface. To hold y constant and allow x to vary, we pass the plane $y = y_1$ through S, intersecting S along the curve AB. We now draw line PQ through P tangent to the curve AB. Then $\partial z/\partial x$ equals the slope of PQ. Similarly, to hold x constant and allow y to vary, we pass the plane $x = x_1$ through S, intersecting S along the curve CD. We now draw line PR through P tangent to the curve CD. Then $\partial z/\partial y$ equals the slope of PR.

EXAMPLE 8.33 The *equation of state* (or *characteristic equation*) of an ideal gas of constant mass is $PV = kT$, where P, V, and T denote absolute pressure, volume, and absolute temperature, respectively, and k is a constant. If V and T can be varied at will, find the partial derivatives of P.

SOLUTION Solving the given equation for P, we have

$$P = \frac{kT}{V} = kTV^{-1}$$

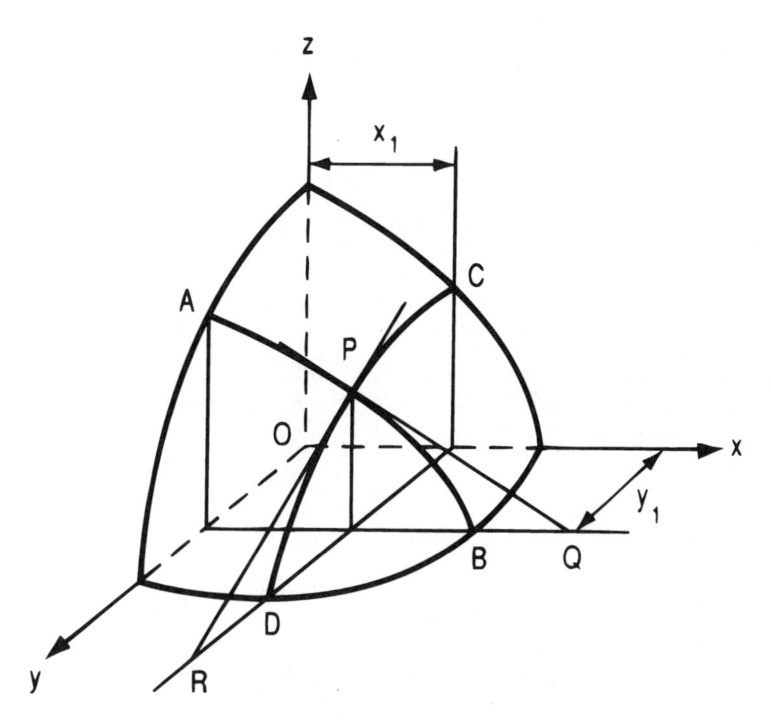

FIGURE 8.17 Geometric interpretation of partial derivatives.

Holding T constant while V varies, we have

$$\frac{\partial P}{\partial V} = -\frac{kT}{V^2}$$

Holding V constant while T varies, we have

$$\frac{\partial P}{\partial T} = \frac{k}{V}$$

8.3.2 Partial Derivatives of Higher Order

The definition of a derivative of higher order in Art. 8.1.5 can be extended to partial derivatives. The notation is illustrated by the following forms:

$$\frac{\partial^2 z}{\partial x^2} = \frac{\partial}{\partial x}\left(\frac{\partial z}{\partial x}\right) \qquad \frac{\partial^2 z}{\partial y\,\partial x} = \frac{\partial}{\partial y}\left(\frac{\partial z}{\partial x}\right)$$

EXAMPLE 8.34 If $z = \sin x - \cos xy$, what is $\partial^2 z/\partial x^2$ and $\partial^2 z/(\partial y\,\partial x)$?

SOLUTION Treating y as a constant and x as a variable, we obtain

$$\frac{\partial z}{\partial x} = \cos x + y \sin xy \tag{c}$$

Repeating this operation, we obtain

$$\frac{\partial^2 z}{\partial x^2} = -\sin x + y^2 \cos xy$$

Now returning to Eq. (c) and treating x as a constant and y as a variable, we obtain

$$\frac{\partial^2 z}{\partial y\,\partial z} = y\frac{\partial}{\partial y}(\sin xy) + \sin xy\frac{\partial y}{\partial y}$$

$$= xy \cos xy + \sin xy$$

The geometric interpretation of second-order partial derivatives is this: In Fig. 8.17, let m denote the slope of the tangent PQ that is parallel to the xz plane. Then $\partial z/\partial x = m$, and $\partial^2 z/\partial x^2$ is the rate at which m changes as we move along curve AB. On the other hand, $\partial^2 z/(\partial y\,\partial x)$ is the rate at which m changes as we move along curve CD.

We shall now illustrate two applications of partial derivatives.

8.3.3 Related Rates of Change with Three Variables

Consider that x and y vary with time but are independent of each other, again let z denote a function of x and y, and let t denote elapsed time. By an extension of Eq.

(8.8), we obtain the following as the rate of change of z:

$$\frac{dz}{dt} = \frac{\partial z}{\partial x}\frac{dx}{dt} + \frac{\partial z}{\partial y}\frac{dy}{dt} \tag{8.12}$$

EXAMPLE 8.35 A right circular cone undergoes a continuous change in size. At a given instant, the following data apply: The altitude is 40 units, and it is increasing at the rate of 3 units/s; the radius of the base is 6 units, and it is decreasing at the rate of 0.8 units/s. Find the instantaneous rate of change of the volume, and devise an approximate test.

SOLUTION The volume of a right circular cone is

$$V = \frac{\pi r^2 h}{3}$$

where h, r, and V denote the altitude, the radius of the base, and the volume, respectively. The partial derivatives are as follows:

$$\frac{\partial V}{\partial h} = \frac{\pi r^2}{3} = \frac{\pi \times 6^2}{3} = 12\pi$$

$$\frac{\partial V}{\partial r} = \frac{2\pi rh}{3} = \frac{2\pi \times 6 \times 40}{3} = 160\pi$$

$$\frac{dV}{dt} = \frac{\partial V}{\partial h}\frac{dh}{dt} + \frac{\partial V}{\partial r}\frac{dr}{dt} = 12\pi \times 3 + 160\pi(-0.8) = -92\pi \text{ units}^3/\text{s}$$

The negative result signifies that the volume is decreasing at the given instant.

This result can be tested approximately by a procedure similar to that in Example 8.25. Assume that the rates of change remain constant for 1/1000 s. The initial volume is

$$V = \frac{\pi \times 6^2 \times 40}{3} = 480\pi \text{ units}^3$$

At the end of this time interval, the volume would be

$$V = \frac{\pi(5.9992)^2(40.003)}{3} = 479.9080\pi \text{ units}^3$$

The decrease in volume would be 0.0920π units^3 to four significant figures. Multiplying this value by 1000, we obtain 92π units^3.

8.3.4 Maxima and Minima with Three Variables

Again let S denote the surface having the equation $z = f(x,y)$. If S contains a summit or nadir, the plane that is tangent to S at that point is parallel to the xy plane. It follows that $\partial z/\partial x = \partial z/\partial y = 0$ at a summit or nadir.

Conversely, if both partial derivatives are 0 at a given point, that point is a summit, a nadir, or a saddle point, which is a point that is a summit in one direction and a nadir in the perpendicular direction. Usually, the specific character of the point is apparent from the conditions of the problem. If doubt exists, we can compute the second derivatives and apply the criterion presented in Art. 8.2.2.

EXAMPLE 8.36 Plane Q has the equation $3x - 2y - z = 5$. Identify the point M on this plane that lies closest to the point $P(3,7,-2)$. Then compute the distance between plane Q and point P.

SOLUTION We first rewrite the given equation in the form $z = 3x - 2y - 5$. Let $R(x,y,z)$ denote an arbitrary point on plane Q, and let $w = (RP)^2$. Applying Eq. (7.23) for the distance between two points, we obtain

$$w = (x-3)^2 + (y-7)^2 + [3x - 2y - 5 - (-2)]^2$$
$$= 10x^2 + 5y^2 - 12xy - 24x - 2y + 67$$

To minimize w, we set

$$\frac{\partial w}{\partial x} = 20x - 12y - 24 = 0$$

$$\frac{\partial w}{\partial y} = -12x + 10y - 2 = 0$$

The solution of this system of simultaneous equations is $x = 33/7$ and $y = 41/7$. Then $z = -18/7$. These are the coordinates of point M.

The distance between plane Q and point P is that between points M and P. By Eq. (7.23),

$$(MP)^2 = \left(\frac{33}{7} - 3\right)^2 + \left(\frac{41}{7} - 7\right)^2 + \left(-\frac{18}{7} + 2\right)^2 \qquad MP = \frac{4\sqrt{14}}{7}$$

EXAMPLE 8.37 Establish the values of x and y at which z is maximum or minimum, where

$$z = 2x^2 - 3xy + 4y^2 + 23x - 46y + 8 \qquad (d)$$

Compute the corresponding value of z and state whether it is maximum or minimum.

SOLUTION

$$\frac{\partial z}{\partial x} = 4x - 3y + 23 \qquad \frac{\partial z}{\partial y} = -3x + 8y - 46$$

Setting each partial derivative equal to 0 and solving the resulting system of simultaneous equations, we obtain $x = -2$, $y = 5$. Substituting this set of values in Eq. (*d*), we obtain $z = -130$.

The computed value of z is *minimum*, and this statement can be proved by three alternative methods. The first method applies simple logic. It is evident from Eq. (*d*) that if we hold y constant and allow x to increase without limit, z becomes infinitely large. Thus, there is no upper bound to the value of z. The second method entails the second partial derivatives, which are as follows:

$$\frac{\partial^2 z}{\partial x^2} = 4 \qquad \frac{\partial^2 z}{\partial y^2} = 8$$

Since both second derivatives are positive, it follows that the computed value of z is minimum. The third method consists of assigning slightly different values to x and y and computing the corresponding value of z. Arbitrarily set $x = -1.9$ and $y = 5.1$. The corresponding value of z is -129.97, and it follows that the value -130 is the minimum value assumed by z.

CHAPTER 9

INTEGRAL CALCULUS

9.1 INTEGRATION

9.1.1 Meaning of Integration

Let y be a function of x. In the process of differentiation, we are given the expression for y in terms of x, and we are required to find the derivative dy/dx. The derivative yields the differential dy, which is defined in Art. 8.1.7. The reverse process is called *integration*. In integration, we are given the expression for the differential dy, and we must construct the expression for y.

The expression for y that results from integration is called the *integral*. To signify that integration is to be performed, we place the symbol $\int$ before the differential. This symbol, which is a distorted S, is called the *integral sign*. The function that follows the integral sign is the *integrand*. For example, in

$$\int \sin^2 x \cos x \, dx$$

the integrand is $\sin^2 x \cos x$.

As an illustration, let $y = x^3$. We have the following:

$$\frac{dy}{dx} = 3x^2 \qquad dy = 3x^2 \, dx$$

Therefore, we may tentatively write

$$\int 3x^2 \, dx = x^3$$

Differentiation is a many-one process. For example, consider the function

$$y = x^2 + C$$

where C is a constant. Then $dy/dx = 2x$ and $dy = 2x \, dx$ regardless of the value of C. It follows that there is an infinite number of functions having $2x \, dx$ as their differentials. Recognizing this fact, we must append a *constant of integration* C to our integral. Thus, returning to our illustrative case, we have

$$\int 3x^2 \, dx = x^3 + C$$

Where an integral contains a constant of integration that can be assigned an arbitrary value, the integral is said to be *indefinite*. However, if we are given a set of simultaneous values of x and y, we can evaluate this constant.

Table A.2 in the appendix presents a set of integrals (with the constant of integration omitted). The integrals are numbered for reference purposes. Integrals (3) and (4) are of fundamental importance. The following are illustrative applications:

$$\int 8 \sin x \, dx = 8 \int \sin x \, dx$$

$$\int (3x^4 - 9x) \, dx = \int 3x^4 \, dx - \int 9x \, dx$$

EXAMPLE 9.1 Evaluate the following:

$$y = \int \frac{dx}{9x^2 - 12x + 29}$$

SOLUTION Refer to Integrals (27). In the present case,

$$a = 9 \qquad b = -12 \qquad c = 29 \qquad 4ac - b^2 = 900 = 30^2$$

Therefore, Integral (27*a*) applies, and the present integral is

$$y = \frac{2}{30} \tan^{-1} \left(\frac{18x - 12}{30} \right) + C = \frac{1}{15} \tan^{-1} (0.6x - 0.4) + C$$

EXAMPLE 9.2 Evaluate the following and verify the result:

$$y = \int \cos^2 6x \, dx$$

SOLUTION Refer to Integral (14). We set $u = 6x$; then $du = 6\,dx$, and $dx = (du)/6$. Now we rewrite the integrand to conform to Integral (14) and proceed to integrate. The result is as follows:

$$y = \int \cos^2 u \frac{du}{6} = \frac{1}{6}\left(\frac{u}{2} + \frac{\sin 2u}{4}\right) + C = \frac{1}{6}\left(\frac{6x}{2} + \frac{\sin 12x}{4}\right) + C$$

$$= \frac{x}{2} + \frac{\sin 12x}{24} + C$$

This result is verified by formulating the expression for dy. Applying Derivatives (3) and (16) and then Eq. (6.15), we obtain the following:

$$\frac{dy}{dx} = \frac{1}{2} + \frac{12 \cos 12x}{24} = \frac{1 + \cos 12x}{2} = \frac{1 + (2\cos^2 6x - 1)}{2} = \cos^2 6x$$

Then

$$dy = \cos^2 6x \, dx$$

and our expression for y is confirmed.

We shall now illustrate the application of integration in problem solving.

EXAMPLE 9.3 A mothball of spherical shape loses mass at a rate directly proportional to its surface area. It has been observed that the diameter diminishes to 60 percent of its original value in 140 days. Compute the following: the time T_1 required for the diameter to diminish to 45 percent of its original value; the time T_2 required for the mass to diminish to 30 percent of its original value; the time T_3 required for the surface area to diminish to 80 percent of its original value. The volume V and surface area A of a sphere are as follows:

$$V = \frac{\pi D^3}{6} \qquad A = \pi D^2$$

where D denotes the diameter.

SOLUTION Let m denote the mass of the sphere, ρ denote the density of the material (i.e., the ratio of mass to volume), and t denote elapsed time. We shall append the subscript 0 to m, D, and A to denote the initial value of the respective variable.

The mass and diameter of the sphere diminish simultaneously. Therefore, this problem encompasses two interrelated time rates: dm/dt and dD/dt. Let k denote the constant of proportionality in these rates. Since the observed information pertains to dD/dt, the solution centers about the expression for this derivative. The steps in the solution are as follows:

1. **Formulate the expression for *dm/dt* in terms of *k*.**

$$\frac{dm}{dt} = -kA = -k\pi D^2 \tag{a}$$

2. **Establish the relationship between *dD/dt* and *dm/dt*.**

$$m = \rho V = \frac{\pi \rho D^3}{6} \tag{b}$$

$$\frac{dm}{dD} = \frac{3\pi\rho D^2}{6} = \frac{\pi\rho D^2}{2} \qquad \text{and} \qquad dD = \frac{2}{\pi\rho D^2}\,dm$$

Then

$$\frac{dD}{dt} = \frac{2}{\pi\rho D^2}\frac{dm}{dt} \tag{c}$$

3. **Formulate the expression for *dD/dt* in terms of *k*.** Replacing dm/dt in Eq. (*c*) with its expression in Eq. (*a*), we obtain

$$\frac{dD}{dt} = -\frac{2k}{\rho} \tag{d}$$

4. **Formulate the expression for *D* in terms of *k*.** Recasting Eq. (*d*) in differential form and then integrating, we obtain

$$dD = -\frac{2k}{\rho}\,dt \qquad D = -\frac{2kt}{\rho} + C$$

Since $D = D_0$ when $t = 0$, $C = D_0$, and the last equation becomes

$$D = D_0 - \frac{2kt}{\rho} \tag{e}$$

5. **Applying the given data, formulate the final expression for *D* and then for *t*.** In Eq. (*e*), we set $t = 140$ days and $D = 0.60D_0$ to obtain

$$0.6D_0 = D_0 - \frac{2k \times 140}{\rho} = D_0 - \frac{280k}{\rho}$$

Then
$$\frac{k}{\rho} = \frac{0.4D_0}{280} = \frac{D_0}{700}$$

Substituting this expression in Eq. (*e*), we obtain

$$D = D_0\left(1 - \frac{t}{350}\right)$$

and
$$t = 350\left(1 - \frac{D}{D_0}\right) \tag{f}$$

6. **Compute T_1.** Setting $D = 0.45D_0$ in Eq. (*f*), we obtain $T_1 = 350(0.55) = 192.5$ days.
7. **Compute T_2.** Equation (*b*) reveals that D is directly proportional to the cube root of m. Applying this relationship and setting $m = 0.3m_0$, we obtain the following:

$$\frac{D}{D_0} = \left(\frac{m}{m_0}\right)^{1/3} = \left(\frac{0.3m_0}{m_0}\right)^{1/3} = 0.6694$$

Substituting this value in Eq. (*f*), we obtain $T_2 = 350(0.3306) = 115.7$ days.

8. **Compute T_3.** The diameter is directly proportional to the square root of the surface area. Setting $A = 0.8A_0$, we obtain

$$\frac{D}{D_0} = (0.80)^{1/2} = 0.8944$$

Equation (*f*) yields $T_3 = 350(0.1056) = 37.0$ days.

9.1.2 Exponentially Varying Quantities

A vast number of variables that arise in engineering, science, medicine, and economics are characterized by the following relationship:

$$\frac{dy}{dx} = ky \tag{g}$$

where k is a constant that can be positive or negative. Expressed verbally, the variable y changes at a rate directly proportional to its own magnitude. We shall investigate such variables.

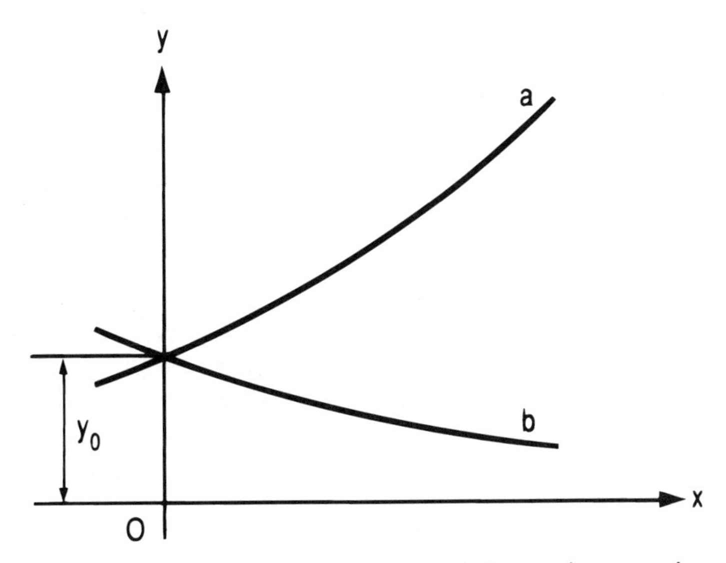

FIGURE 9.1 Graph of an exponentially varying quantity.

Let y_0 denote the value of y when $x = 0$. In Fig. 9.1, the variation of y is represented by curve a when k is positive and by curve b when k is negative. At every point, the slope of the tangent to the curve is directly proportional to the ordinate. By rearranging Eq. (g) and then applying Integral (7) with $a = 1$ and $b = 0$, we obtain the following:

$$\frac{dy}{y} = k\,dx \qquad \ln y = kx + C$$

Applying the definition of the natural logarithm presented in Art. 1.8, we now obtain this result:

$$y = e^{kx+C} = e^{kx}\,e^{C}$$

Since $y = y_0$ when $x = 0$, it follows that $e^C = y_0$, and we have

$$y = y_0\,e^{kx} \tag{9.1}$$

Thus, y varies *exponentially* with x.

The mass of a radioactive substance illustrates a quantity that diminishes exponentially with time because the mass diminishes continuously at a rate directly proportional to its own magnitude. By convention, scientists express the rate of radioactive disintegration of a given substance by means of its *half-life period*. This is the amount of time required for the mass to diminish to one-half of its original value.

EXAMPLE 9.4 A certain radioactive substance has a half-life period of 1200 years. How long does it take for the mass of the substance to diminish to 80 percent of its original value?

SOLUTION Let y denote the mass and t denote elapsed time. In Eq. (9.1), we replace x with t and proceed to solve for t, in this manner:

$$e^{kt} = \frac{y}{y_0} \qquad kt = \ln \frac{y}{y_0}$$

$$t = \frac{\ln (y/y_0)}{k} \qquad (h)$$

Let T denote the required time. Applying Eq. (h), we have

$$\frac{T}{1200} = \frac{\ln 0.8}{\ln 0.5} \qquad T = 386.3 \text{ years}$$

EXAMPLE 9.5 The disintegration of a certain radioactive substance is characterized by the equation $dm/dt = -0.08m$, where m denotes the mass and t denotes elapsed time in hours. When $t = 0$, $m = 35$. What is the elapsed time when $m = 10$?

SOLUTION Applying Eq. (h) and replacing y with m, we obtain

$$t = \frac{\ln (10/35)}{-0.08} = 15.66 \text{ h}$$

EXAMPLE 9.6 When a slab of porous material is exposed to air, it dries at a rate directly proportional to its moisture content. Under specified weather conditions, a slab loses 40 percent of its moisture in 0.5 h. Compute the time required for the slab to lose 75 percent of its moisture under the same conditions.

SOLUTION The amount of moisture in the slab diminishes at a rate directly proportional to its own magnitude. Let T denote the required time. Proceeding as in Example 9.4 and taking the proportion of moisture *that remains*, we obtain the following:

$$\frac{T}{0.5} = \frac{\ln 0.25}{\ln 0.60} \qquad T = 1.357 \text{ h}$$

9.1.3 Calculation of a Plane Area

In Art. 8.1.8, we considered that a line starts at the y axis in Fig. 8.2 and then moves to the right while remaining parallel to the y axis. As the line moves, it generates an area bounded by the curve, the x axis, the y axis, and its present position. This area is denoted by A. Equation (8.7) yields

$$A = \int y \, dx \qquad (9.2)$$

EXAMPLE 9.7 A curve has the equation $y = 3x^2$. Compute the area that lies between the x axis and the curve and that extends from the origin to the line $x = 4$.

SOLUTION

$$A = \int 3x^2 \, dx = x^3 + C$$

Since $A = 0$ when $x = 0$, it follows that $C = 0$. Replacing x with 4, we obtain $A = 64$. The unit of A is the product of the units of x and y.

9.1.4 Integrating Differentials of Higher Order

In Art. 8.1.5, we defined the nth derivative of y with respect to x. It follows from the definition that

$$\int \frac{d^n y}{dx^n}\,dx = \frac{d^{n-1}y}{dx^{n-1}} + C$$

EXAMPLE 9.8 The slope of the tangent to a curve varies at the rate of $30x - 16$ with respect to x, and the curve contains the points $(2,-17)$ and $(6,771)$. Formulate the equation of the curve.

SOLUTION As stated in Art. 8.1.5, the rate at which the slope of the tangent changes equals the second derivative of y with respect to x. Then

$$\frac{d^2y}{dx^2} = 30x - 16$$

To solve for y, we must integrate twice. Therefore, two constants of integration will arise. When we multiply both sides of the foregoing equation by dx and integrate, we obtain

$$\frac{dy}{dx} = \frac{30x^2}{2} - 16x + C_1 = 15x^2 - 16x + C_1$$

Repeating the foregoing operation, we now obtain

$$y = \frac{15x^3}{3} - \frac{16x^2}{2} + C_1x + C_2 = 5x^3 - 8x^2 + C_1x + C_2$$

Replacing x with 2 and y with -17, we obtain

$$2C_1 + C_2 = -25$$

Now replacing x with 6 and y with 771, we obtain

$$6C_1 + C_2 = -21$$

The solution of this system of simultaneous equations is $C_1 = 1$ and $C_2 = -27$. Therefore, the equation of the curve is

$$y = 5x^3 - 8x^2 + x - 27$$

9.1.5 The Definite Integral

With reference to Fig. 9.2, assume that we wish to find the area bounded by the curve, the x axis, and the straight lines having the equations $x = a$ and $x = b$. Let A denote this area. The procedure is as follows: Applying Eq. (9.2), compute the area A_a from the y axis to $x = a$, compute the area A_b from the y axis to $x = b$, and then set $A = A_b - A_a$. The values a and b are termed the *limits of integration*, a being the lower limit and b being the upper limit. Symbolically, the area is represented by placing the limits of integration after the integral sign, with the upper limit above

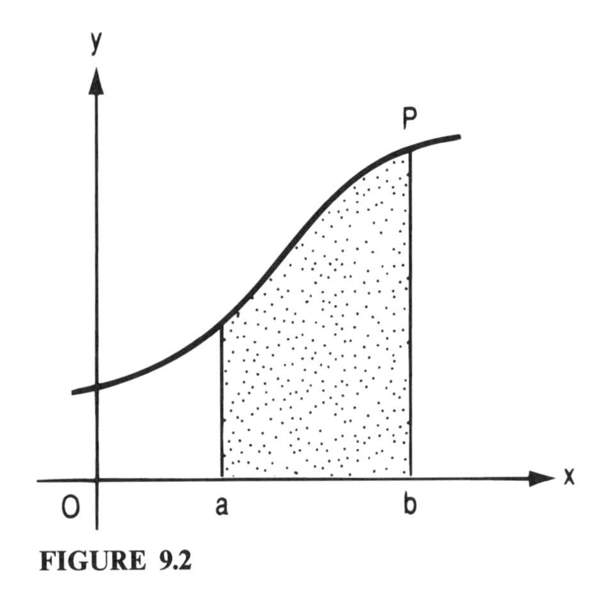

FIGURE 9.2

the line and the lower limit below the line, in this manner:

$$A = \int_a^b y \, dx \tag{i}$$

If the expression that results from integration is a single term, it is followed by a closing bracket, and the limits of integration are placed after the bracket. If the expression contains several terms, the entire expression is enclosed in brackets.

When A_a is subtracted from A_b, the constant of integration vanishes. Consequently, an integral that contains limits of integration is termed a *definite integral.* A definite integral can be regarded as the increase in the value of the integral across the interval between $x = a$ and $x = b$.

EXAMPLE 9.9 A curve has the equation $y = x^2 + x$. Compute the area bounded by the curve, the x axis, and the lines $x = 2$ and $x = 5$.

SOLUTION

$$A = \int_2^5 (x^2 + x) \, dx = \left[\frac{x^3}{3} + \frac{x^2}{2}\right]_2^5$$

The subtraction can be performed within each term, giving

$$A = \frac{5^3 - 2^3}{3} + \frac{5^2 - 2^2}{2} = 49.5$$

EXAMPLE 9.10 A curve has the equation $y = 3x^2 - 75$. Find the total amount of area (in absolute value) bounded by the curve, the x axis, and the lines $x = -2$ and $x = 6$.

SOLUTION Care must be exercised where the curve lies partly above and partly below the x axis, as straight integration between the given limits will yield simply the *net* area. Setting $y = 0$, we obtain $x = \pm 5$, and we find that the curve lies below the x axis in the interval $-5 < x < 5$ and above the x axis elsewhere. Therefore, we must divide the given interval into two subintervals: from $x = -2$ to $x = 5$, and from $x = 5$ to $x = 6$.

$$\int (3x^2 - 75)\, dx = x^3 - 75x$$

$$\begin{aligned} A &= -[x^3 - 75x]_{-2}^{5} + [x^3 - 75x]_5^6 \\ &= -[5^3 - (-2)^3] + 75[5 - (-2)] + (6^3 - 5^3) - 75(6 - 5) \\ &= 408 \end{aligned}$$

A more detailed investigation yields the following information: From $x = -2$ to $x = 5$, the area is -392; from $x = 5$ to $x = 6$, the area is 16. Therefore, straight integration between $x = -2$ and $x = 6$ would have yielded the misleading result $A = -376$.

9.1.6 Interpretation of Integral as an Area

Since the amount of area between a curve and the x axis is an integral, we can invert our point of view to conceive that a given integral is equal to an area. To illustrate this concept, assume that a point moves in a straight line. Let s, v, and t denote distance traversed, velocity, and elapsed time, respectively. Then

$$v = \frac{ds}{dt} \qquad \therefore\ ds = v\, dt \qquad \Delta s = \int v\, dt$$

If we draw a diagram in which t is plotted on the horizontal axis and v is plotted on the vertical axis, the area between the resulting curve and the horizontal axis equals the distance traversed. As we shall find, enormous benefits accrue by viewing an integral as the area under a curve.

9.1.7 Improper Integrals

A definite integral having the limits a and b is said to be *improper* if either a or b is infinite or if both are infinite, or if the integrand becomes infinite within the interval a to b.

EXAMPLE 9.11 Compute the area that lies between the curve $y = 1/x^2$ and the x axis and that extends indefinitely to the right of the line $x = a$, where a is positive.

SOLUTION The curve is plotted in Fig. 8.3. The limits of integration are a and infinity.

$$A = \int_a^\infty x^{-2}\, dx = -\frac{1}{x}\bigg]_a^\infty = \frac{1}{a}$$

Thus, although the area is of infinite extent, it has a finite value, and the integral is said to *converge* to this value.

9.1.8 The Fundamental Theorem

We have found that an area can be viewed as an integral, and we shall now view the area as a sum. These alternative ways of viewing an area lead to a principle that is of vast significance.

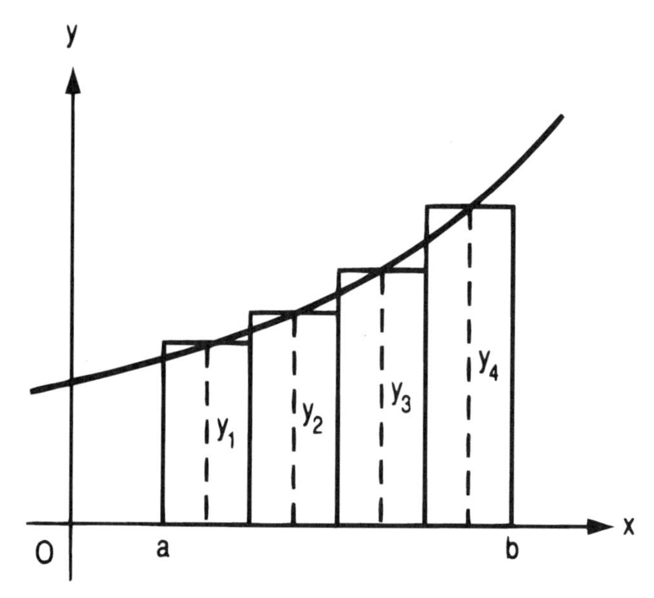

FIGURE 9.3 Development of the fundamental theorem.

Referring to Fig. 9.3, consider again that we wish to evaluate the area bounded by the curve, the x axis, and the lines $x = a$ and $x = b$. We proceed in this manner: Divide the interval $x = a$ to $x = b$ into n parts, and number the parts from left to right. (The parts need not be of uniform width.) Let Δx_i denote the width of the ith part. Draw lines parallel to the y axis at the boundaries. Now select an arbitrary point on the curve within each part, and draw a line through this point parallel to the x axis. Let y_i denote the ordinate to the arbitrary point in the ith part. We have thus constructed n rectangles. The total area of these rectangles is

$$\sum_{i=1}^{n} y_i \, \Delta x_i$$

Now consider that n is increased by making the widths of the parts progressively smaller. As the widths approach 0 (and n becomes infinitely large), the total area of the rectangles approaches the area between the curve and the x axis as a limit. Therefore, applying our previous expression for the area, we obtain

$$\lim_{\Delta x_i \to 0} \sum y_i \, \Delta x_i = \int_a^b y \, dx \tag{9.3}$$

This relationship, which is known as the *fundamental theorem* of integral calculus, enables us to view the definite integral as a sum. We shall now illustrate how the theorem is applied.

EXAMPLE 9.12 A force acts on a body, causing the body to move in the direction of the force. The magnitude of the force varies with the displacement of the body, in this manner:

$$F = 0.4 + 1.7s + 0.13s^2$$

where F is the force in newtons (N) and s is the displacement of the body in meters (m). Compute the work performed in displacing the body 12 m from its original position.

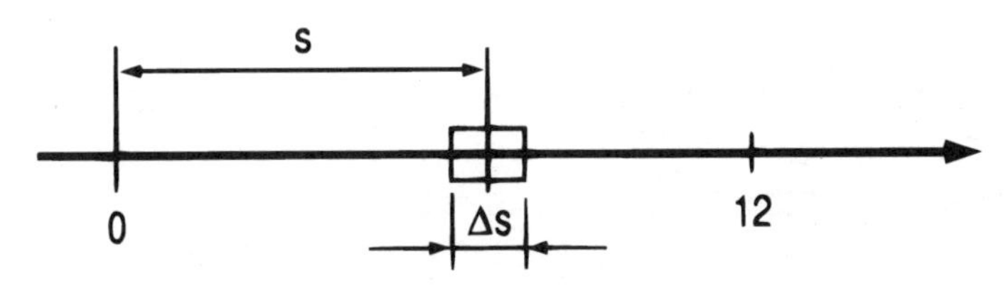

FIGURE 9.4

SOLUTION If the force acting on a body remains constant, the work performed by the force is the product of the force and the displacement of the body in the direction of the force. In the present case, the force varies.

Referring to Fig. 9.4, consider that the distance through which the body moves is divided into small parts. Let Δs denote the length of one part, and let s denote the displacement to the center of this part. Assume that the force remains constant across the part, and let ΔW denote the work performed in moving the body through this part. Then

$$\Delta W = (0.4 + 1.7s + 0.13s^2)\,\Delta s$$

To find the total work W performed, we sum the work performed across all the parts, allow the parts to become progressively smaller, and apply Eq. (9.3). Then

$$W = \int_0^{12} (0.4 + 1.7s + 0.13s^2)\,ds$$

$$= \left[0.4s + \frac{1.7s^2}{2} + \frac{0.13s^3}{3}\right]_0^{12} = 202.08$$

The unit of work is the Joule (J), where 1 J = 1 N · m.

In our subsequent applications of the fundamental theorem, we shall replace the symbol Δ with the letter d. Thus, ΔW becomes dW.

EXAMPLE 9.13 The force F required to deform a spring an amount δ is given by

$$F = 3.5\delta + 0.016\delta^2$$

where F is in newtons and δ is in millimeters. How much work is performed when the elongation of the spring changes from 50 mm to 80 mm?

SOLUTION If F is directly proportional to δ, the spring is described as *linear*. In the present case, the spring is nonlinear.

When the spring has an elongation of δ, the force acting on the spring has the magnitude given by the foregoing equation. We proceed as in Example 9.12. If the force remained constant, the work performed in increasing the deformation by an amount $d\delta$ would be

$$dW = F\,d\delta = (3.5\delta + 0.016\delta^2)\,d\delta$$

Then

$$W = \int_{50}^{80} (3.5\delta + 0.016\delta^2)\,d\delta = \left[\frac{3.5\delta^2}{2} + \frac{0.016\delta^3}{3}\right]_{50}^{80}$$

$$= \frac{3.5(80^2 - 50^2)}{2} + \frac{0.016(80^3 - 50^3)}{3}$$

$$= 8889 \text{ N} \cdot \text{mm} = 8.889 \text{ N} \cdot \text{m} = 8.889 \text{ J}$$

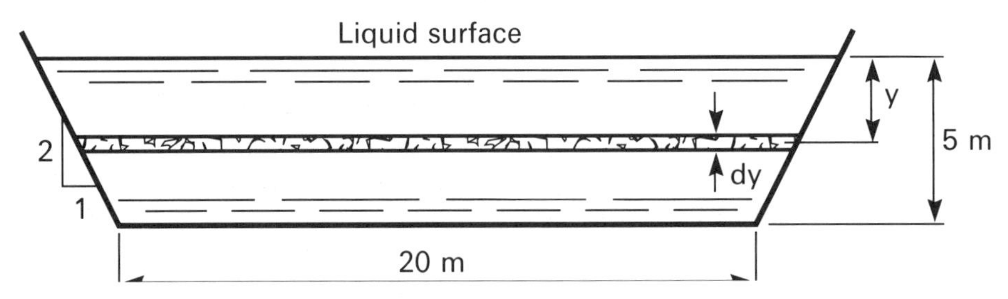

FIGURE 9.5 Evaluating the force on a dam.

EXAMPLE 9.14 A dam has a vertical face in the shape of an isosceles trapezoid, as shown in Fig. 9.5. If the specific weight of water is 9800 N/m^3, what is the total hydrostatic force on the face of the dam when the liquid surface is 5 m above the base?

SOLUTION The *specific weight* of a homogeneous body is the ratio of its weight to its volume. The force acting on a unit area of a body is termed *pressure*. The pressure p that a liquid exerts at a distance y below the liquid surface is $y\gamma$, where γ denotes specific weight.

Let a denote the width of the dam at a given elevation. In Fig. 9.5, $a = 20 + 2(0.5) \times 5 = 25$ m at the liquid surface. Consider the force dF acting on the infinitesimally thin horizontal strip shown shaded. Let y denote the distance from the liquid surface to the center of this strip, dy denote the thickness of the strip, and dA denote the area of the strip. At the center, $a = 25 - y$. Assume tentatively that the pressure remains constant across the depth of the strip. The force dF acting on the strip is $dF = p\ dA$. Now, $p = y\gamma$ and $dA = a\ dy$. Then $dF = y\gamma a\ dy$, or

$$dF = \gamma(25 - y)y\ dy = \gamma(25y - y^2)\ dy$$

and the total force F is

$$F = \gamma \int_0^5 (25y - y^2)\ dy = 9800\left[\frac{25y^2}{2} - \frac{y^3}{3}\right]_0^5$$

$$= 2.654 \times 10^6 \text{ N} = 2.654 \text{ MN}$$

9.1.9 Volume of a Solid of Revolution

We shall now illustrate another important application of the fundamental theorem. If a solid object is generated by revolving a portion of a plane about an axis that lies within the plane, the object is referred to as a *solid of revolution*. The plane surface that is revolved is called the *section* of the solid. In the subsequent material, we shall apply the following relationships: Let r, C, and A denote, respectively, the radius, circumference, and area of a circle. Then $C = 2\pi r$ and $A = \pi r^2$.

In Fig. 9.6, PQ is a continuous curve, and lines AP and BQ are parallel to the y axis. Consider that the region $ABQP$ is revolved about the x axis through an angle of 2π. We wish to evaluate the volume of the solid thus formed.

We divide the region into n parts by means of lines parallel to the y axis. One part, of width Δx, is shown dotted. We select an arbitrary point E within this part, draw ER parallel to the y axis, where R lies on the curve, and draw a line through R parallel to the x axis, thereby forming a rectangle. Let $y = ER$. When this rectangle revolves about the x axis, line ER generates a circle of area πy^2, and the rectangle

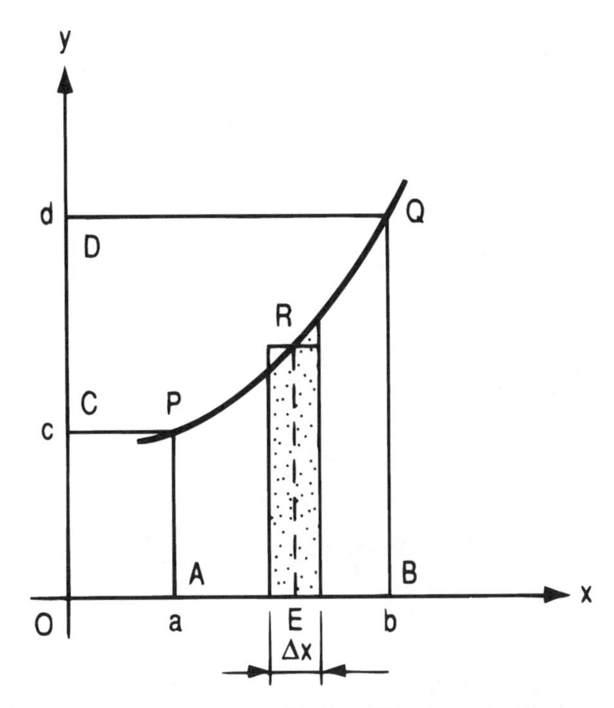

FIGURE 9.6 Generating a solid of revolution.

generates a right circular cylinder, or *disk*, of volume $\pi y^2 \Delta x$. By summing the volumes of these disks, allowing n to become infinite, and applying the fundamental theorem, we obtain

$$V = \pi \int_a^b y^2 \, dx \tag{9.4a}$$

Similarly, if the region $CPQD$ is revolved about the y axis, the volume of the solid of revolution thus generated is

$$V = \pi \int_c^d x^2 \, dy \tag{9.4b}$$

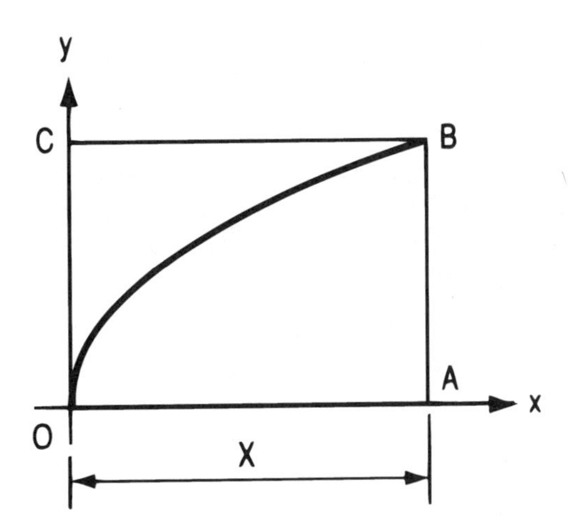

FIGURE 9.7

EXAMPLE 9.15 The curve in Fig. 9.7 has the equation $y = x^{1/2}$, line AB is parallel to the y axis, and $OA = X$. Compute the volume generated when the region OAB is revolved about the x axis.

SOLUTION Since $y^2 = x$, Eq. (9.4a) becomes

$$V = \pi \int_0^X x \, dx = \pi \left. \frac{x^2}{2} \right]_0^X = \frac{\pi X^2}{2}$$

Now consider that the region $ABQP$ in Fig. 9.6 is revolved about the y axis through an angle of 2π. In this instance, the "solid" of revolution is a hollow body. Let $OE = x$. Point E describes a circle of circumference $2\pi x$, and the dotted rectangle forms a *cylindrical shell* of thickness Δx. If this thickness is infinitesimal, we may set the volume of this shell equal to $2\pi xy\ \Delta x$. By summing the volumes of these shells, allowing n to become infinite, and applying the fundamental theorem, we obtain

$$V = 2\pi \int_a^b xy\ dx \tag{9.5}$$

EXAMPLE 9.16 Compute the volume generated when the region OAB in Fig. 9.7 is revolved about the y axis.

SOLUTION Applying Eq. (9.5), we obtain

$$V = 2\pi \int_0^X x(x^{1/2})\ dx = 2\pi \int_0^X x^{3/2}\ dx = 2\pi \frac{X^{5/2}}{5/2} = \frac{4\pi X^{5/2}}{5}$$

9.2 RECTILINEAR (STRAIGHT-LINE) MOTION

9.2.1 Definitions, Notation, and Units

The basic notational system for motion is as follows:

t = elapsed time

s = distance traversed

v = velocity

a = acceleration (time rate of change of velocity)

If the unit of time is the second (s) and the unit of distance is the meter (m), the unit of velocity is meters per second (m/s), and the unit of acceleration is meters per second per second [(m/s)/s], or m/s^2.

A quantity that has magnitude only is termed a *scalar*. Thus, the temperature of a body and the number of electrons in an atom are scalars. A quantity that has both magnitude and direction is termed a *vector*. Thus, a force, a velocity, and an acceleration are vectors. If a moving point undergoes a change in the magnitude or direction of its velocity, or both, it has acceleration. The direction of a vector has two characteristics: *inclination* (e.g., 28° with the horizontal) and *sense* (e.g., southwestward).

9.2.2 General Rectilinear Motion

The motion of a point in a straight line is called *rectilinear*. Since the inclination of motion remains constant, we may view velocity and acceleration as scalars within the present context, and we shall use the term *direction* to denote the *sense* of the motion.

The motion of a point in a plane can be depicted by means of three diagrams, in each of which time is plotted on the horizontal axis. The quantity plotted on the vertical axis is as follows: in the *s-t diagram*, distance traversed; in the *v-t diagram*, velocity; in the *a-t diagram*, acceleration.

Let Δs and Δv denote the change in distance and velocity, respectively, during a given time interval. In the following material, we shall refer to the slope of the tangent to a curve as simply the slope of the curve. The basic relationships are as follows:

$$v = \frac{ds}{dt}$$

and v equals the slope of the *s-t* diagram.

$$a = \frac{dv}{dt} = \frac{d^2s}{dt^2}$$

and a equals the slope of the *v-t* diagram.

$$\Delta s = \int v \, dt$$

and Δs equals the area between the *v-t* diagram and the t axis.

$$\Delta v = \int a \, dt$$

and Δv equals the area between the *a-t* diagram and the t axis.

In Examples 9.17 through 9.19, the unit of time is the second.

EXAMPLE 9.17 A point moves along a straight line, and the equation of its motion is $s = 3t^2 - 0.04t^3$. Compute the following: the initial velocity and acceleration; the velocity and acceleration when $t = 5$ s; the time at which the point reverses its direction; the maximum displacement of the point in the positive direction.

SOLUTION

$$v = \frac{ds}{dt} = 6t - 0.12t^2$$

$$a = \frac{dv}{dt} = 6 - 0.24t$$

Setting $t = 0$, we obtain $v = 0$ and $a = 6$ units/s^2. Setting $t = 5$ s, we obtain $v = 27$ units/s and $a = 4.8$ units/s^2.

The initial velocity of the point is 0. The acceleration is positive at first but eventually changes to negative. Therefore, the velocity is initially positive and eventually negative. At the instant the point is reversing its direction, $v = 0$. We therefore set

$$6t - 0.12t^2 = 0 \qquad \text{or} \qquad t(6 - 0.12t) = 0$$

The two solutions are $t = 0$, which is irrelevant in the present instance, and $t = 50$ s. Thus, when $t < 50$, the point moves in the positive direction; thereafter, the point moves in the negative direction.

The maximum displacement s occurs when $t = 50$ s, and its value is 2500 units.

EXAMPLE 9.18 A point moving in a straight line has an initial velocity of 12 units/s, and its acceleration is $a = 3t + 2$. Find the distance traversed when $t = 3$ s.

SOLUTION Starting with the given equation for a and integrating twice, we obtain the following results:

$$a = \frac{dv}{dt} = 3t + 2$$

$$v = \frac{3t^2}{2} + 2t + C_1 = 1.5t^2 + 2t + C_1$$

$$s = \frac{1.5t^3}{3} + \frac{2t^2}{2} + C_1 t + C_2 = 0.5t^3 + t^2 + C_1 t + C_2$$

Since $s = 0$ when $t = 0$, $C_2 = 0$. Since $v = 12$ when $t = 0$, $C_1 = 12$. Therefore, the equation for s is

$$s = 0.5t^3 + t^2 + 12t$$

Setting $t = 3$, we obtain $s = 58.5$ units.

EXAMPLE 9.19 A particle moves in a straight line, and its motion is characterized by the equation $v = 6t^2 - 54$. Initially, the particle is 5 units from a reference point. What is the distance between the particle and the reference point at the instant the particle reverses the direction of its motion? What is the acceleration of the particle at that instant?

SOLUTION Let s denote the distance between the particle and the reference point. Starting with the given equation for v, we obtain the following results:

$$s = \int v\,dt = \frac{6t^3}{3} - 54t + C = 2t^3 - 54t + 5$$

$$a = \frac{dv}{dt} = 12t$$

The direction of motion changes when $v = 0$, and this condition occurs when $t = 3$ s. The corresponding value of s is -103 units, and the corresponding value of a is 36 units/s^2.

9.2.3 Rectilinear Motion with Constant Acceleration

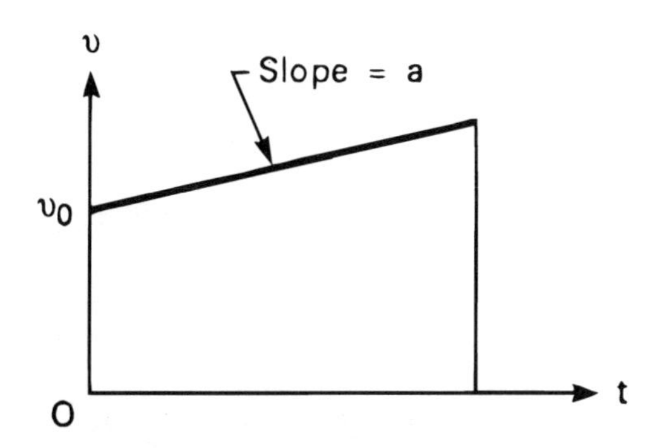

FIGURE 9.8 Velocity-time diagram with constant acceleration.

A form of rectilinear motion that arises frequently in practice is that in which the acceleration is constant. Let s_0 denote the initial displacement of the moving point from a reference point and v_0 denote the initial velocity, and let s and v denote the displacement and velocity, respectively, at time t. The v-t diagram is a trapezoid, as shown in Fig. 9.8, and the s-t diagram is a parabolic arc. The relationships are as follows:

$$v = v_0 + at \tag{9.6}$$

Setting $s - s_0$ equal to the area of the trapezoid, we obtain

$$s = s_0 + \left(\frac{v_0 + v}{2}\right)t \tag{9.7}$$

Integration of Eq. (9.6) yields

$$s = s_0 + v_0 t + \frac{at^2}{2} \tag{9.8}$$

By solving Eq. (9.6) for t and substituting the resulting expression in Eq. (9.7), we obtain

$$v^2 = v_0^2 + 2a(s - s_0) \tag{9.9}$$

The advantage of applying Eq. (9.9) is that it circumvents the calculation of time.

EXAMPLE 9.20 A particle moves upward along an inclined plane with a constant deceleration of 8 units/s^2. What must be its velocity at the base if the particle is to rise a distance of precisely 49 units as measured along the plane? How long will it take the particle to traverse this distance? Verify the results.

SOLUTION When the particle reaches its summit, its velocity is 0. Applying Eq. (9.9) with $v = 0$, $s_0 = 0$, $a = -8$, and $s = 49$, we obtain

$$0 = v_0^2 + 2(-8)49 \qquad v_0 = 28 \text{ units/s}$$

Applying Eq. (9.6) with this value, we obtain $t = 3.5$ s.

By substituting our calculated values in Eq. (9.8), we find that $s = 49$ units, and our results are thus confirmed.

9.2.4 Rotation

Consider that a line L rotates about the origin in the counterclockwise direction. Let θ denote the angle between L and the positive side of the x axis. Let ω and α

denote, respectively, the angular velocity and angular acceleration of L.

Mathematically, the rotation of a line is exactly analogous to the rectilinear motion of a point, and the equations of rectilinear motion can be adapted to rotation simply by replacing s with θ, v with ω, and a with α.

9.3 MULTIPLE INTEGRATION

9.3.1 Meaning of Multiple Integration

Consider again that z is a function of x and y and that x and y can vary independently of each other. Multiple (or repeated) integration is the reverse of partial differentiation. In multiple integration, we integrate in steps, allowing only one independent quantity to vary in each step.

The integral that results is termed a *double integral* if the number of independent variables is two, and a *triple integral* if this number is three. The general notation of double integration is as follows:

$$\int_c^d \int_a^b f(x,y)\, dy\, dx$$

In step 1, we hold x constant, let y vary, and integrate the expression $f(x,y)\, dy$ between the limits a and b. Call I the resulting integral. In step 2, we let x vary, and integrate the expression $I\, dx$ between the limits c and d. Thus, the sequence in which the quantities vary corresponds to the order in which the differentials are recorded. A limit of integration of a variable may be a function of the value of the succeeding variable.

EXAMPLE 9.21 Evaluate the following:

$$\int_2^4 \int_x^{2x} \int_0^6 (xy + z^2)\, dz\, dy\, dx$$

SOLUTION Holding x and y constant and allowing z to vary, we obtain

$$\int_0^6 (xy + z^2)\, dz = \left[xyz + \frac{z^3}{3} \right]_0^6$$

$$= 6xy + 72$$

Now holding x constant and allowing y to vary, we obtain

$$\int_x^{2x} (6xy + 72)\, dy = [3xy^2 + 72y]_x^{2x}$$

$$= 9x^3 + 72x$$

Now allowing x to vary, we obtain

$$\int_2^4 (9x^3 + 72x)\,dx = \left[\frac{9x^4}{4} + 36x^2\right]_2^4$$

$$= 972$$

9.3.2 Volume of a Solid

We shall apply multiple integration to compute the volume of a general type of solid. We shall assume for simplicity that the base of the solid lies in the xy plane; the top face of the solid can be a plane or a curved surface.

In Fig. 9.9, consider that we pass through the solid closely spaced planes that are parallel to the xy and yz planes, thereby dividing the solid into strips parallel to the z axis. One such strip is shown dotted. Let Δx and Δy denote the dimensions of the section of this strip, as shown, and let z be the height of the strip at its centerline. The volume ΔV of this strip is approximately $z\ \Delta x\ \Delta y$. If we sum the volumes of these strips, allow the number of strips to become infinite, and apply the fundamental theorem, we obtain the following as the total volume V of the solid:

$$V = \iint z\,dy\,dx \tag{9.10}$$

EXAMPLE 9.22 A solid has its base in the region of the xy plane bounded by the parabola $y = x^2 - 8$ and the straight line $y = 2x$. The top face is the plane $z = x + 5$. Compute the volume of the solid.

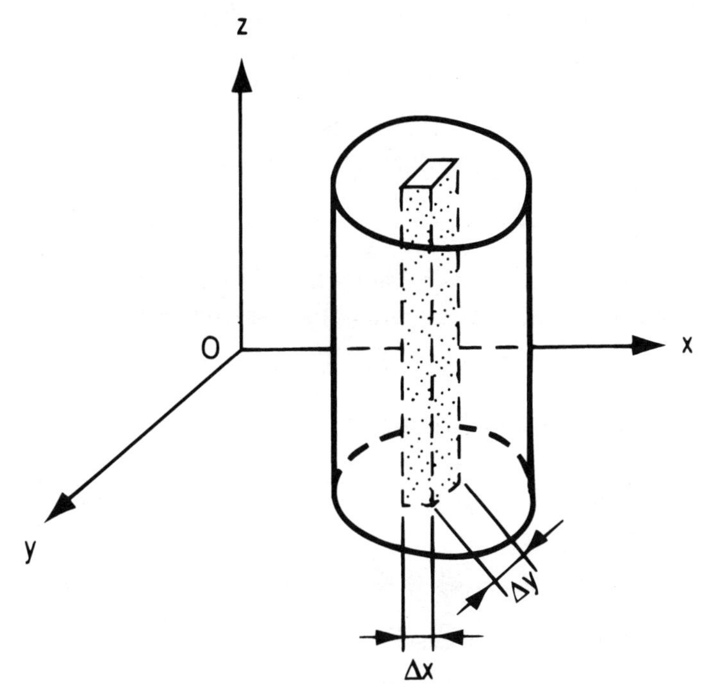

FIGURE 9.9 Division of solid into strips.

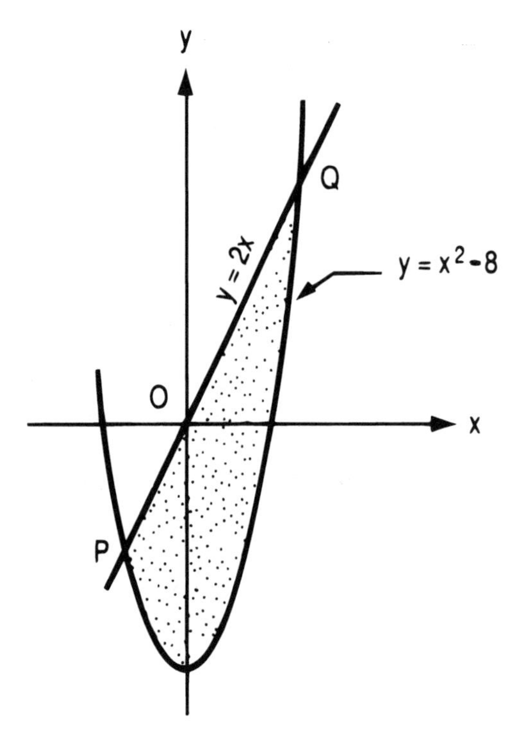

FIGURE 9.10 Base of solid.

SOLUTION The base of the solid is shown in Fig. 9.10. The range of y is from $x^2 - 8$ to $2x$. To find the range of x, we must locate the intersection points P and Q. We therefore set $x^2 - 8 = 2x$. The solution of this equation is $x = -2$, $x = 4$. Then

$$V = \int_{-2}^{4} \int_{x^2-8}^{2x} (x + 5)\, dy\, dx$$

The first integration yields

$$[(x + 5)y]_{x^2-8}^{2x} = -x^3 - 3x^2 + 18x + 40$$

The second integration yields

$$V = \left[-\frac{x^4}{4} - x^3 + 9x^2 + 40x \right]_{-2}^{4} = 216$$

The first integration, in which x was held constant and y varied, gave us the volume of a strip that is parallel to the yz plane and extends completely through the solid. The second integration gave us the sum of the volumes of all such strips.

9.4 POWER SERIES AND FOURIER SERIES

9.4.1 General Form of Power Series

In Art. 1.15, we defined a power series, and we shall now consider power series that are infinite. If y is a function of x, the series has the form

$$A_0 + A_1 y + A_2 y^2 + \cdots + A_n y^n + \cdots$$

where the A's are all constant. Whether this series is convergent or divergent depends on the value of y, and thus on the value assigned to x.

9.4.2 Taylor and Maclaurin Series

Assume that a function of x can be expressed as the sum of a power series in $x - a$, where a is a constant. Then

$$f(x) = A_0 + A_1(x - a) + A_2(x - a)^2 + A_3(x - a)^3 + \cdots$$

By differentiating this equation successively and setting $x = a$, we obtain the values of the coefficients. Substituting these values, we obtain the following:

$$f(x) = f(a) + f'(a)(x - a) + \frac{f''(a)}{2!}(x - a)^2 + \frac{f'''(a)}{3!}(x - a)^3 + \cdots + \frac{f^n(a)}{n!}(x - a)^n + \cdots \tag{9.11}$$

The series at the right is termed a *Taylor series*, and $f(x)$ is said to be *expanded* in a Taylor series.

If we now set $a = 0$, Eq. (9.11) reduces to

$$f(x) = f(0) + f'(0)x + \frac{f''(0)}{2!}x^2 + \frac{f'''(0)}{3!}x^3 + \cdots + \frac{f^n(0)}{n!}x^n + \cdots \tag{9.12}$$

The series at the right is termed a *Maclaurin series*, and it is a special form of the Taylor series. The Taylor and Maclaurin series are of extreme importance, for they provide a means of evaluating many functions.

EXAMPLE 9.23 Expand $\ln(1 + x)$ in powers of x.

SOLUTION The procedure consists of obtaining the successive derivatives of the given function, then setting $x = 0$ in each expression, and substituting the results in Eq. (9.12).

$$f(x) = \ln(1 + x) \qquad f(0) = \ln 1 = 0$$

$$f'(x) = (1 + x)^{-1} \qquad f'(0) = 1$$

$$f''(x) = -(1 + x)^{-2} \qquad f''(0) = -1$$

$$f'''(x) = 2!(1 + x)^{-3} \qquad f'''(0) = 2!$$

$$f^{iv}(x) = -3!(1 + x)^{-4} \qquad f^{iv}(0) = -3!$$

The pattern is now clear, and we see that

$$f^n(0) = (-1)^{n-1}[(n - 1)!]$$

Substituting these results in Eq. (9.12), we obtain

$$\ln(1 + x) = x - \frac{x^2}{2} + \frac{x^3}{3} - \frac{x^4}{4} + \cdots$$

EXAMPLE 9.24 Expand $\sin x$ and $\cos x$ in powers of x.

SOLUTION For $\sin x$, we have the following:

$$f(x) = \sin x \qquad f(0) = 0$$

$$f'(x) = \cos x \qquad f'(0) = 1$$

$$f''(x) = -\sin x \qquad f''(0) = 0$$

$$f'''(x) = -\cos x \qquad f'''(0) = -1$$

$$f^{\text{iv}}(x) = \sin x \qquad f^{\text{iv}}(0) = 0$$

$$f^{\text{v}}(x) = \cos x \qquad f^{\text{v}}(0) = 1$$

The pattern is now clear. Substituting these results in Eq. (9.12), we obtain

$$\sin x = x - \frac{x^3}{3!} + \frac{x^5}{5!} - \frac{x^7}{7!} + \cdots \tag{j}$$

For cos x, we have the following:

$$f(x) = \cos x \qquad f(0) = 1$$

$$f'(x) = -\sin x \qquad f'(0) = 0$$

$$f''(x) = -\cos x \qquad f''(0) = -1$$

$$f'''(x) = \sin x \qquad f'''(0) = 0$$

$$f^{\text{iv}}(x) = \cos x \qquad f^{\text{iv}}(0) = 1$$

$$f^{\text{v}}(x) = -\sin x \qquad f^{\text{v}}(0) = 0$$

The pattern is again clear. Substituting these results in Eq. (9.12), we obtain

$$\cos x = 1 - \frac{x^2}{2!} + \frac{x^4}{4!} - \frac{x^6}{6!} + \cdots \tag{k}$$

EXAMPLE 9.25 Write the first five terms in the Maclaurin expansion of $(\cos^2 x)/x$.

SOLUTION The simplest procedure consists of expanding the function $\cos^2 x$ and then dividing each term in the series by x. We shall apply the relationship $\sin 2x = 2 \sin x \cos x$.

$$f(x) = \cos^2 x \qquad f(0) = 1$$

$$f'(x) = -2 \cos x \sin x$$

$$= -\sin 2x \qquad f'(0) = 0$$

$$f''(x) = -2 \cos 2x \qquad f''(0) = -2$$

$$f'''(x) = 4 \sin 2x \qquad f'''(0) = 0$$

$$f^{\text{iv}}(x) = 8 \cos 2x \qquad f^{\text{iv}}(0) = 8$$

$$f^{\text{v}}(x) = -16 \sin 2x \qquad f^{\text{v}}(0) = 0$$

$$f^{\text{vi}}(x) = -32 \cos 2x \qquad f^{\text{vi}}(0) = -32$$

Substituting these results in Eq. (9.12), we obtain the following:

$$\cos^2 x = 1 - \frac{2}{2!}x^2 + \frac{8}{4!}x^4 - \frac{32}{6!}x^6 + \frac{128}{8!}x^8 - \cdots$$

We now perform the division within the coefficients and divide each term by x to obtain

$$\frac{\cos^2 x}{x} = \frac{1}{x} - x + \frac{x^3}{3} - \frac{2x^5}{45} + \frac{x^7}{315} - \cdots$$

9.4.3 Euler's Formula

In Art. 1.20, we developed Eq. (1.20) by applying the binomial theorem. We now recognize that this equation expresses e^x in the form of a power series. In Art. 1.9, we stated that i denotes $\sqrt{-1}$, and it is the unit of imaginary numbers. If we now replace x with ix in Eq. (1.20) and compare the resulting series with those in Eqs. (j) and (k) of Example 9.24, we find that

$$e^{ix} = \cos x + i \sin x \tag{9.13}$$

This relationship, which is known as *Euler's formula*, has numerous applications. For example, in electrical engineering, where quantities vary sinusoidally with time, it enables us to express these quantities in terms of $e^{i\theta}$ rather than in terms of $\sin \theta$ or $\cos \theta$. Thus, Euler's formula provides an alternative mode of expression.

If we now replace x with $-ix$ in Eq. (1.20) and compare the resulting series for e^{-ix} with that for e^{ix}, we obtain the following:

$$\frac{e^{ix} + e^{-ix}}{2} = \cos x \tag{9.14a}$$

$$\frac{e^{ix} - e^{-ix}}{2i} = \sin x \tag{9.14b}$$

9.4.4 Fourier Series for Periodic Functions

A *periodic function* is one whose values recur in cycles. The *period* or *interval* of the function is the amount by which the independent variable increases in value in one cycle. For example, $\sin x$ and $\cos x$ are periodic functions having a period of 2π. The transition from one cycle to another may be smooth, as in the case of $\sin x$ and $\cos x$, or discontinuous, as in the case of the periodic function represented in Fig. 9.11.

We shall term $\sin nx$ and $\cos nx$, where n can have any integral value, the *basic* periodic functions. Each of these has a period of $2\pi/n$. This question suggests itself: Is it possible to view a general periodic function as a composite of an infinite number of basic periodic functions of steadily diminishing period, plus a constant? Fourier demonstrated that such is the case if the given funcion satisfies certain continuity requirements.

Let $F(t)$ denote a periodic function of t. Let τ denote its period, and let $\omega = 2\pi/\tau$.

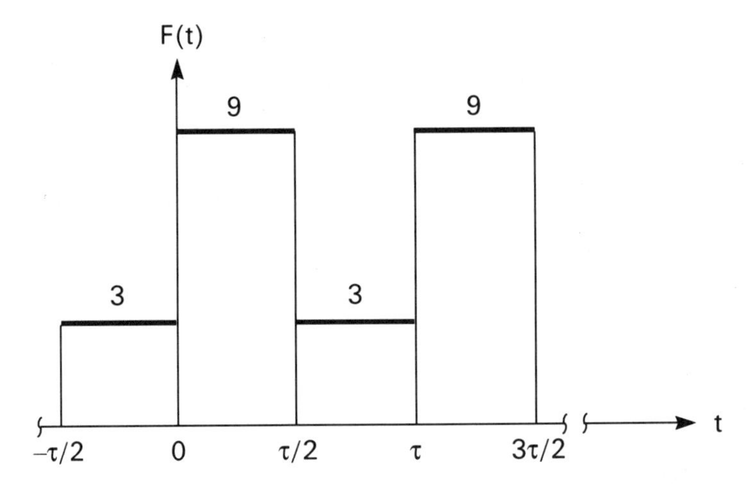

FIGURE 9.11 Periodic function.

Then $F(t)$ can be expressed in the form of the following infinite series:

$$F(t) = \frac{a_0}{2} + a_1 \cos \omega t + a_2 \cos 2\omega t + \cdots + a_n \cos n\omega t + \cdots$$

$$+ b_1 \sin \omega t + b_2 \sin 2\omega t + \cdots + b_n \sin n\omega t + \cdots \qquad (9.15)$$

The series at the right is termed a *Fourier series*, and the function is said to be *expanded* in its Fourier series.

The coefficients associated with a given function can be evaluated by this procedure: If we multiply both sides of Eq. (9.15) by $\cos n\omega t\, dt$ and then integrate between the limits 0 and τ, all terms at the right vanish except that which was originally $a_n \cos n\omega t$. If we now multiply both sides of Eq. (9.15) by $\sin n\omega t\, dt$ and then integrate between the limits 0 and τ, all terms at the right vanish except that which was originally $b_n \sin n\omega t$. By solving the equations that result from integration for a_n and b_n, we obtain the following results:

$$a_n = \frac{2}{\tau} \int_0^\tau F(t) \cos n\omega t\, dt \qquad (9.16a)$$

$$b_n = \frac{2}{\tau} \int_0^\tau F(t) \sin n\omega t\, dt \qquad (9.16b)$$

As an illustration, refer to the periodic function represented in Fig. 9.11. If $\tau = 2\pi$, $\omega = 1$. Equations (9.16) yield these results:

$$a_0 = 12 \qquad a_n = 0 \qquad b_n = -\frac{1}{n\pi}(6 \cos n\pi - 6)$$

Equation (9.15) now yields

$$F(t) = 6 + \frac{12}{\pi}\left(\sin t + \frac{\sin 3t}{3} + \frac{\sin 5t}{5} + \cdots\right)$$

9.5 DIFFERENTIAL EQUATIONS

9.5.1 Definitions

In our present discussion, we shall consider x and y to be the independent and dependent variables, respectively. If we are given a value of x and the corresponding value of y or of a derivative of y, these values are referred to as *boundary values.*

A *differential equation* is one that contains derivatives or differentials. Thus, Example 9.8 required the solution of a differential equation, but of an elementary type. An *ordinary* differential equation is one that does not contain partial derivatives. The *order* of a differential equation is the highest order of a derivative in the equation. For example, if the equation contains d^4y/dx^4 but no higher derivative, the equation is of the fourth order.

The *degree* of a term in a differential equation is found by adding the exponents of y and its derivatives. For example, the term $x(d^3y/dx^3)$ is of degree 1, and the term $y(dy/dx)^2$ is of degree 3. A *linear* differential equation is one in which each term that contains y or its derivatives is of degree 1.

The general form of a linear equation of order n is

$$a_0 \frac{d^ny}{dx^n} + a_1 \frac{d^{n-1}y}{dx^{n-1}} + a_2 \frac{d^{n-2}y}{dx^{n-2}} + \cdots + a_{n-1} \frac{dy}{dx} + a_n y = f(x)$$

where the coefficients are functions of x or constants.

Consider that we are given the expression for dy/dx, and let r denote a constant. If the replacement of x and y with rx and ry, respectively, leaves dy/dx unchanged, the differential equation is described as *homogeneous.* For example, the following equation is homogeneous:

$$\frac{dy}{dx} = \frac{-5x^2 + 4xy + 9y^2}{2x^2 + 6xy}$$

An expression for y in terms of x, in either explicit or implicit form, is a *solution* of a given differential equation if that expression contains no derivatives and if it satisfies the given equation. In the absence of any boundary values, a differential equation generally has an infinite set of solutions. A *particular solution* is a solution within this set; a *general solution* is one that encompasses the entire set. The general solution contains arbitrary constants that result from integration, and the number of these constants equals the order of the differential equation.

As an illustration, consider the second-order differential equation

$$\frac{d^2y}{dx^2} + 2\frac{dy}{dx} - 15y = 0$$

A particular solution of this equation is $y = e^{3x}$, as we can verify in this manner:

$$\frac{dy}{dx} = 3e^{3x} \qquad \frac{d^2y}{dx^2} = 9e^{3x}$$

Substituting in the given equation, we obtain

$$9e^{3x} + 6e^{3x} - 15e^{3x} = 0$$

and the solution is thus confirmed. However, another particular solution is

$y = e^{-5x}$, and the general solution is $y = C_1 e^{3x} + C_2 e^{-5x}$, where C_1 and C_2 are the arbitrary constants.

The subject of differential equations is very broad, and we shall confine our coverage to the types of equations that are likely to appear in the Fundamentals of Engineering Examination.

9.5.2 Type 1 Equations

An equation of type 1 is of the first order, and it can be restructured in such manner that all x terms become associated with dx and all y terms become associated with dy. The general form is

$$F(x)\,dx + G(y)\,dy = 0$$

The equation can be solved directly by integration.

EXAMPLE 9.26 Solve the equation

$$\frac{dy}{dx} = \frac{x^3 + 6}{8xy}$$

SOLUTION To arrange the variables in the prescribed manner, we multiply both sides of the equation by $8y\,dx$, giving

$$8y\,dy = \left(x^2 + \frac{6}{x}\right) dx$$

Integration yields

$$4y^2 = \frac{x^3}{3} + 6 \ln x + C = \frac{x^3}{3} + \ln x^6 + C$$

where C is the arbitrary constant.

9.5.3 Type 2 Equations

An equation of type 2 is of the first order, it is linear, and it can be expressed in the general form

$$\frac{dy}{dx} + Py = Q$$

where P and Q are either functions of x alone or constants. The left side of the equation can be transformed to the derivative of a product if both sides of the equation are multiplied by an *integrating factor* R that has the value

$$R = e^{\int P\,dx}$$

The differential equation is then solved by integrating both sides.

In solving a type 2 equation, it is often necessary to apply the identity $e^{\ln A} = A$. This relationship stems from the definition of $\ln A$: the power to which e must be raised to obtain A.

EXAMPLE 9.27 Solve the equation

$$\frac{dy}{dx} = \frac{2x^5 + 3y}{x}$$

SOLUTION To cast the equation in standard form, we replace the fraction at the right with $2x^4 + 3y/x$ and then transpose. The result is

$$\frac{dy}{dx} - \frac{3}{x} y = 2x^4 \tag{l}$$

Then $P = -3/x$, and we have

$$\int P \, dx = -3 \ln x = \ln x^{-3}$$

$$R = e^{\int P \, dx} = x^{-3}$$

We now multiply both sides of Eq. (l) by R to obtain

$$x^{-3} \frac{dy}{dx} - 3x^{-4} y = 2x$$

The expression at the left is the derivative of $x^{-3}y$, and integration yields

$$x^{-3} y = \int 2x \, dx = x^2 + C$$

We now multiply by x^3, and the solution of the differential equation becomes

$$y = x^5 + Cx^3$$

If a differential equation is linear with respect to x rather than y, the same method of solution can be followed by simply reversing the roles of the two variables.

9.5.4 Type 3 Equations

A type 3 equation is a special form of a type 2 equation in which P is a constant and $Q = 0$. We shall write it in this general form:

$$\frac{dy}{dx} + ay = 0$$

where a is a real constant. The formula for solving a type 2 equation yields this general solution of a type 3 equation:

$$y = Ce^{-ax} \tag{9.17}$$

where C is the arbitrary constant.

EXAMPLE 9.28 Solve the following differential equation and verify the result:

$$\frac{dy}{dx} - 5y = 0$$

SOLUTION Applying Eq. (9.17) with $a = -5$, we obtain the solution $y = Ce^{5x}$, where C is the arbitrary constant.

We now have $dy/dx = 5Ce^{5x}$. When the expressions for y and dy/dx are substituted in the given equation, the result is 0, and our solution is thus confirmed.

9.5.5 Type 4 Equations

An equation of type 4 is of the second order, and it has the same characteristics as an equation of type 3 in all other respects. The general form is

$$\frac{d^2y}{dx^2} + a_1 \frac{dy}{dx} + a_2 y = 0$$

where a_1 and a_2 are real constants. Associated with this equation is an *auxiliary* (or *characteristic*) *equation* that is formed in this manner: Replace y with 1, replace dy/dx with p, and replace d^2y/dx^2 with p^2. Thus, the auxiliary equation is

$$p^2 + a_1 p + a_2 = 0$$

The solution of this auxiliary equation yields the solution of the differential equation, and three possibilities exist. We shall investigate each one in turn. In the following material, C_1 and C_2 denote the arbitrary constants.

Case 1. The roots of the auxiliary equation are real and distinct. Let r and s denote the roots. The solution of the differential equation is

$$y = C_1 e^{rx} + C_2 e^{sx} \tag{9.18}$$

EXAMPLE 9.29 Solve the following differential equation and verify the result:

$$\frac{d^2y}{dx^2} - 3\frac{dy}{dx} - 28y = 0$$

SOLUTION The auxiliary equation is $p^2 - 3p - 28 = 0$, and the roots of this equation are -4 and 7. Since these roots are real and distinct, the differential equation falls within the scope of Case 1, and Eq. (9.18) yields the solution

$$y = C_1 e^{-4x} + C_2 e^{7x}$$

Then $$\frac{dy}{dx} = -4C_1 e^{-4x} + 7C_2 e^{7x} \qquad \frac{d^2y}{dx^2} = 16C_1 e^{-4x} + 49C_2 e^{7x}$$

When we substitute the expressions for y and its derivatives in the given differential equation, we find that the equation is satisfied. Our result is thus confirmed.

EXAMPLE 9.30 A point moves in a straight line according to the equation

$$\frac{d^2x}{dt^2} - 7\frac{dx}{dt} + 12x = 0$$

where x denotes the displacement of the point from a reference point and t denotes elapsed time in seconds. When $t = 0$, the displacement is 2 units and the acceleration is -3 units/s^2. Compute the velocity when $t = 0.5$ s.

SOLUTION The auxiliary equation is $p^2 - 7p + 12 = 0$, and the roots are 3 and 4. Therefore, the solution of the differential equation is

$$x = C_1 e^{3t} + C_2 e^{4t}$$

Then $$v = \frac{dx}{dt} = 3C_1 e^{3t} + 4C_2 e^{4t}$$

$$a = \frac{d^2x}{dt^2} = 9C_1 e^{3t} + 16C_2 e^{4t}$$

Setting $t = 0$ and applying the given set of values, we have

$$x = C_1 + C_2 = 2$$

$$a = 9C_1 + 16C_2 = -3$$

The solution of this system of simultaneous equations is $C_1 = 5$ and $C_2 = -3$. Therefore, when $t = 0.5$,

$$v = 3 \times 5e^{1.5} + 4(-3)e^2 = -34.91 \text{ units/s}$$

Case 2. The roots of the auxiliary equation are identical. Let r denote the roots. The solution of the differential equation is

$$y = (C_1 x + C_2)e^{rx} \tag{9.19}$$

EXAMPLE 9.31 Solve the differential equation

$$\frac{d^2y}{dx^2} + 8\frac{dy}{dx} + 16y = 0$$

SOLUTION The auxiliary equation is $p^2 + 8p + 16 = 0$, and the roots of this equation are both -4. Therefore, the differential equation falls within the scope of Case 2, and Eq. (9.19) yields the solution

$$y = (C_1 x + C_2)e^{-4x}$$

Case 3. The roots of the auxiliary equation are conjugate complex numbers. Let $\alpha + i\beta$ and $\alpha - i\beta$ denote the roots. The solution of the differential equation is

$$y = e^{\alpha x}(C_1 \cos \beta x + C_2 \sin \beta x) \tag{9.20}$$

EXAMPLE 9.32 Solve the differential equation

$$\frac{d^2y}{dx^2} + 6\frac{dy}{dx} + 73y = 0$$

SOLUTION The auxiliary equation is $p^2 + 6p + 73 = 0$, and the roots of this equation are $-3 + i8$ and $-3 - i8$. Therefore, the differential equation falls within the scope of Case 3, and Eq. (9.20) yields the solution

$$y = e^{-3x}(C_1 \cos 8x + C_2 \sin 8x)$$

9.5.6 The Laplace Transform and Its Application

Where certain boundary conditions are known, the solution of a differential equation can often be expediated by applying a device known as *the Laplace transform.*

Let $y = f(t)$. Associated with $f(t)$ is a function that is denoted by $\mathscr{L}[f(t)]$ or $F(s)$. It is constructed by this formula:

$$\mathscr{L}[f(t)] = \int_0^\infty f(t)e^{-st}\,dt \tag{9.21}$$

The function $\mathscr{L}[f(t)]$ is the Laplace transform of $f(t)$, and $f(t)$ is the *inverse transform* of $\mathscr{L}[f(t)]$. We shall present two illustrations of the transformation.

EXAMPLE 9.33 Construct the Laplace transform of $y = \sin \alpha t$.

SOLUTION

$$\mathscr{L}(\sin \alpha t) = \int_0^\infty e^{-st} \sin \alpha t\,dt$$

Referring to a table of integrals, we find that

$$\mathscr{L}(\sin \alpha t) = \left[\frac{e^{-st}(-s \sin \alpha t - \alpha \cos \alpha t)}{s^2 + \alpha^2}\right]_0^\infty$$

As t becomes infinite, e^{-st} approaches 0. When $t = 0$, $e^{-st} = 1$, $\sin \alpha t = 0$, and $\cos \alpha t = 1$. Therefore,

$$\mathscr{L}(\sin \alpha t) = \frac{0 - (-\alpha)}{s^2 + \alpha^2} = \frac{\alpha}{s^2 + \alpha^2}$$

EXAMPLE 9.34 The notation $u(c)$ denotes a unit step when $t = c$. Specifically, $y = 0$ when $t < c$, and $y = 1$ when $t \geq c$. Construct the Laplace transform of this function.

SOLUTION Since the integral in Eq. (9.21) is 0 when $t < c$, we have

$$\mathscr{L}[u(c)] = \int_c^\infty e^{-st} = \left.\frac{e^{-st}}{-s}\right]_0^\infty = \frac{0 - e^{-cs}}{-s} = \frac{e^{-cs}}{s}$$

A comprehensive table of Laplace transforms has been compiled, and a brief listing of transforms appears in Table A.3 in the appendix. The following laws pertaining to Laplace transforms stem directly from the laws of integration:

$$\mathscr{L}[f(t) + g(t)] = \mathscr{L}[f(t)] + \mathscr{L}[g(t)] \tag{9.22}$$

$$\mathscr{L}[cf(t)] = c\mathscr{L}[f(t)] \tag{9.23}$$

where c is a constant.

EXAMPLE 9.35 Develop the Laplace transform of

$$y = 7t - 3 \cos 2t$$

SOLUTION Referring to Table A.3 and applying Eqs. (9.22) and (9.23), we obtain the following:

$$\mathscr{L}(y) = \frac{7}{s^2} - \frac{3s}{s^2 + 4} = \frac{-3s^3 + 7s^2 + 28}{s^4 + 4s^2}$$

The Laplace transform of the nth derivative of y is as follows:

$$\mathscr{L}\left[\frac{d^n f(t)}{dt^n}\right] = -\frac{d^{n-1}f(0)}{dt^{n-1}} - s\frac{d^{n-2}f(0)}{dt^{n-2}} - s^2\frac{d^{n-3}f(0)}{dt^{n-3}} - \cdots - s^{n-1}f(0) + s^n\mathscr{L}[f(t)] \tag{9.24}$$

where the notation (0) denotes that the term is evaluated with $t = 0$. In the particular cases where $n = 1$ and $n = 2$, Eq. (9.24) reduces to the following (with the order reversed):

$$\mathscr{L}\left[\frac{df(t)}{dt}\right] = s\mathscr{L}[f(t)] - f(0) \tag{9.24a}$$

$$\mathscr{L}\left[\frac{d^2f(t)}{dt^2}\right] = s^2\mathscr{L}[f(t)] - sf(0) - \frac{df(0)}{dt} \tag{9.24b}$$

A differential equation can be solved by replacing each term in the given equation with its Laplace transform, solving the resulting equation for $\mathscr{L}(f(t)]$, and then replacing each term in the last equation with its inverse transform. We shall illustrate the procedure.

EXAMPLE 9.36 Solve the equation

$$\frac{d^2y}{dt^2} + 2\frac{dy}{dt} - 15y = 0$$

where $\qquad y(0) = 0 \qquad$ and $\qquad \frac{dy(0)}{dt} = -24$

Verify the result.

SOLUTION The steps in the solution are as follows:

1. **Replace each term in the given equation with its Laplace transform.** By Eq. (9.24b), the transform of d^2y/dt^2 is

$$s^2\mathscr{L}(y) - 8s - (-24) = s^2\mathscr{L}(y) - 8s + 24$$

By Eq. (9.24*a*), the transform of dy/dt is

$$s\mathscr{L}(y) - 8$$

The transform of y is $\mathscr{L}(y)$, and the transform of 0 is 0. Then

$$s^2\mathscr{L}(y) - 8s + 24 + 2s\mathscr{L}(y) - 16 - 15\mathscr{L}(y) = 0$$

2. **Solve the resulting equation for $\mathscr{L}(y)$.** By combining terms, we obtain

$$(s^2 + 2s - 15)\mathscr{L}(y) - 8s + 8 = 0$$

Then

$$\mathscr{L}(y) = \frac{8s - 8}{s^2 + 2s - 15}$$

3. **Factor the denominator**

$$\mathscr{L}(y) = \frac{8s - 8}{s^2 + 2s - 15} = \frac{8s - 8}{(s + 5)(s - 3)}$$

4. **Decompose the last fraction (i.e., express it as the sum of fractions), making each factor the denominator of an individual fraction**

$$\mathscr{L}(y) = \frac{8s - 8}{(s + 5)(s - 3)} = \frac{A}{s + 5} + \frac{B}{s - 3}$$

where A and B are to be evaluated. By cross-multiplying and equating the resulting expression to the original numerator, we obtain

$$A(s - 3) + B(s + 5) = (A + B)s - 3A + 5B = 8s - 8$$

Then

$$A + B = 8$$

$$-3A + 5B = -8$$

The solution of this system of simultaneous equations is $A = 6$, $B = 2$. Therefore,

$$\mathscr{L}(y) = \frac{6}{s + 5} + \frac{2}{s - 3}$$

5. **Replace each term in the last equation with its inverse transform to obtain the equation for y.** From Table A.3, the inverse transform of $1/(s + \alpha)$ is $e^{-\alpha t}$. Then

$$y = 6e^{-5t} + 2e^{3t}$$

The verification is as follows:

$$\frac{dy}{dt} = -30e^{-5t} + 6e^{3t} \qquad \frac{d^2y}{dt^2} = 150e^{-5t} + 18e^{3t}$$

$$y(0) = 6 + 2 = 8 \qquad \frac{dy(0)}{dt} = -30 + 6 = -24$$

When we substitute the expressions for y, dy/dt, and d^2y/dt^2 in the given differential equation, we find that it is satisfied. Thus, our solution meets all three requirements.

CHAPTER 10

PROPERTIES OF AREAS, VOLUMES, AND MASSES

In engineering analysis, certain expressions pertaining to areas, volumes, and masses recur so frequently that it becomes advantageous to assign names to these expressions and to develop their characteristics. We shall first present the properties of areas, and then we shall extend these properties to volumes and masses.

10.1 PROPERTIES OF AREAS

10.1.1 Statical Moment (First Moment)

Consider that the area in Fig. 10.1 is divided into n elements, one of which is shown darkened. Let A denote the total area and dA the area of an element. Also let y denote the ordinate of the center of the element. We form the product $y\,dA$ of each

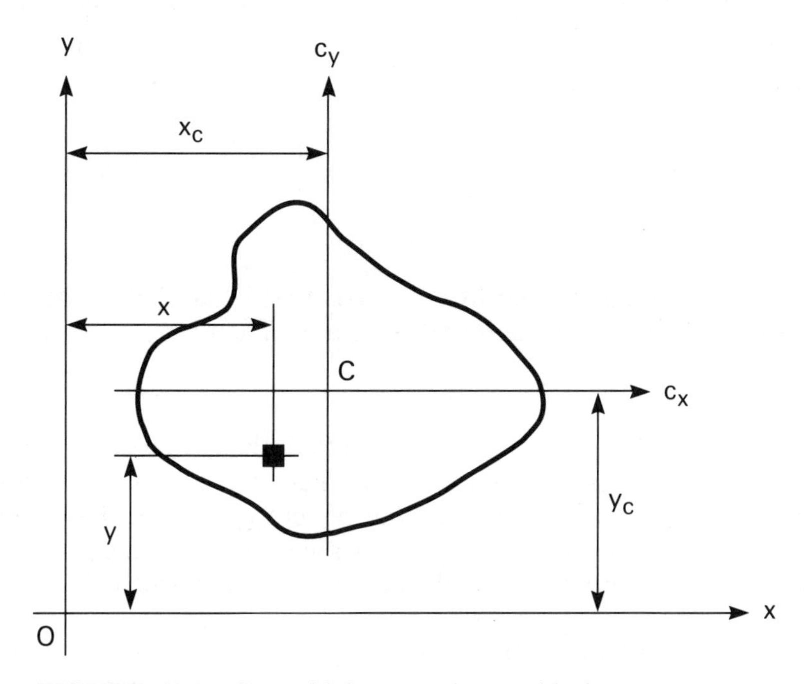

FIGURE 10.1 Centroidal axes and centroid of an area.

element and sum these products. We now make n progressively larger by making the elements progressively smaller. By the fundamental theorem in Art. 9.1.8, we have

$$\lim_{n \to \infty} \sum y \, dA = \int y \, dA$$

where integration is performed across the entire area. This integral is called the *statical moment* (or *first moment*) of the area with respect to the x axis, and it is denoted by M_x. Then

$$M_x = \int y \, dA \tag{10.1a}$$

Analogously, the statical moment of the area with respect to the y axis is

$$M_y = \int x \, dA \tag{10.1b}$$

An area is always considered positive. Therefore, the algebraic sign of $y \, dA$ is identical with the algebraic sign of y, and a statical moment can be positive, negative, or zero. The unit of statical moment is the cube of the unit of length.

The statical moment of an area is the sum of the statical moments of its elements. Therefore, if an area may be considered to be composed of several parts, the statical moment of the area is the sum of the statical moments of its parts. On the other hand, if part of an area is removed, the statical moment of the remaining area is the difference between the statical moment of the original area and that of the part removed.

10.1.2 Centroidal Axes and Centroid

If the statical moment of an area with respect to a given axis is zero, that axis is termed a *centroidal axis*. With reference to Fig. 10.1, let c_x denote the centroidal axis that is parallel to the x axis, y_c denote the distance between this axis and the x axis, and $M_{x,c}$ denote the statical moment of the area with respect to axis c_x. Then

$$M_{x,c} = \int (y - y_c) \, dA = \int y \, dA - y_c \int dA = M_x - Ay_c = 0$$

Then

$$M_x = Ay_c \tag{10.2a}$$

Analogously, let c_y denote the centroidal axis that is parallel to the y axis and x_c denote the distance between this axis and the y axis. Then

$$M_y = Ax_c \tag{10.2b}$$

Now consider that the x and y axes are rotated about O to the positions x' and y', respectively. There are centroidal axes parallel to the x' and y' axes, and it is apparent that an area has an infinite number of centroidal axes. However, it can be demonstrated readily that all centroidal axes are concurrent (i.e., they intersect at the same point), and the point at which they intersect is termed the *centroid* of the area.

A centroidal axis and the centroid of an area can be assigned physical significance in this manner: Consider that we have a thin plate of uniform thickness and homogeneous composition. If this plate is placed in a horizontal position and then placed over a *line support*, the plate will remain at rest if and only if the support lies below a centroidal axis of the section of the plate. Similarly, if the plate is placed in a horizontal position and then placed over a *point support*, the plate will remain at rest if and only if the support lies below the centroid of the section of the plate.

Equations (10.2) can be recast in this manner:

$$y_c = \frac{M_x}{A} \qquad x_c = \frac{M_y}{A} \tag{10.3}$$

Thus, if M_x and M_y are evaluated, these quantities provide a means of locating the centroid of an area.

Assume that an area has an axis of symmetry, and call this the s axis. Corresponding to an element that lies at a distance t from the s axis is an element that lies at a distance $-t$ from the s axis. Thus, the statical moments of the two elements with respect to the s axis nullify each other, and it follows that the statical moment of the area with respect to s is zero. We thus arrive at the following principle:

Theorem 10.1. If an area has an axis of symmetry, this axis is a centroidal axis.

EXAMPLE 10.1 Lines a and b in Fig. 10.2 have the following equations, respectively:

$$2x - 5y = 0 \tag{a}$$

$$3x + 5y = 40 \tag{b}$$

Compute the statical moment with respect to the x axis of the area that lies between these lines and extends from the y axis to their point of intersection P.

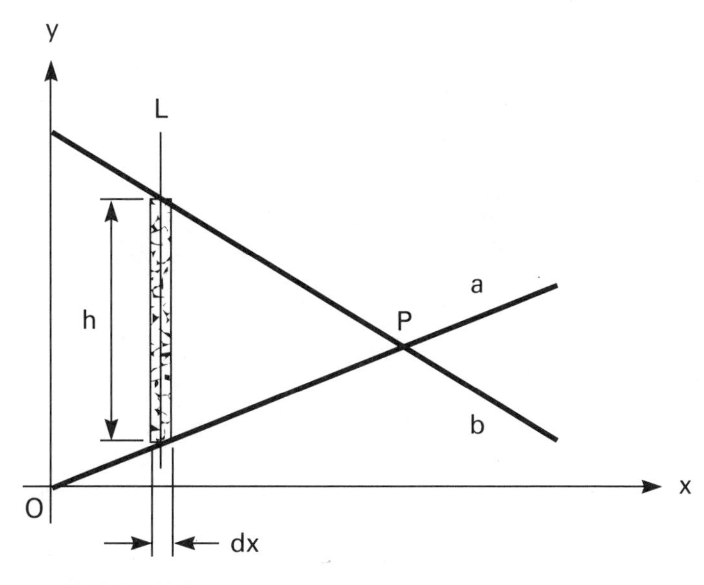

FIGURE 10.2

SOLUTION We shall rewrite the foregoing equations to express y explicitly in terms of x. The results are these:

$$y = 0.4x \tag{a'}$$

$$y = -0.6x + 8 \tag{b'}$$

To obtain the abscissa of P, we set

$$0.4x = -0.6x + 8$$

and we find that $x = 8$ at P.

Method 1. We divide the area into rectangular elements having a width dx parallel to the x axis and dy parallel to the y axis. Therefore, for an individual element, we have the following:

$$dA = dy\,dx \qquad dM_x = y\,dA = y\,dy\,dx$$

To find the statical moment of the total area, we apply double integration. The y values range from $0.4x$ to $-0.6x + 8$, and the x values range from 0 to 8. Then

$$M_x = \int_0^8 \int_{0.4x}^{-0.6x+8} y\,dy\,dx$$

The first integration yields

$$\left[\frac{y^2}{2}\right]_{0.4x}^{-0.6x+8} = \frac{0.36x^2 - 9.6x + 64 - 0.16x^2}{2} = 0.10x^2 - 4.8x + 32$$

and we now have

$$M_x = \int_0^8 (0.10x^2 - 4.8x + 32)\,dx = \left[\frac{0.10x^3}{3} - \frac{4.8x^2}{2} + 32x\right]_0^8$$
$$= 119.47$$

Method 2. In Fig. 10.2, draw line L parallel to the y axis, and then draw a rectangular strip that has L as its centerline and that extends from line a to line b. The strip is shown darkened. Let h denote the height of this strip and dx its thickness. Let y_a and y_b denote the ordinates of the points on L that lie on lines a and b, respectively. For this strip,

$$h = y_b - y_a = -0.6x + 8 - 0.4x = -x + 8$$

$$dA = h\,dx = (-x + 8)\,dx$$

Since this strip is a rectangle, its centerline parallel to the x axis is a centroidal axis. By Eq. (10.2a), the statical moment of this strip with respect to the x axis is

$$dM_x = \left(y_a + \frac{h}{2}\right) dA = \left(0.4x + \frac{-x + 8}{2}\right)(-x + 8)\,dx$$
$$= (-0.1x + 4)(-x + 8)\,dx = (0.1x^2 - 4.8x + 32)\,dx$$

Now consider that we divide the entire region into strips of this type, and let S denote the sum of their statical moments. If we now allow the thickness of the strips to become progressively smaller (and the number of strips progressively larger), S approaches the statical moment of the entire area in the limit. Therefore,

$$M_x = \int_0^8 (0.1x^2 - 4.8x + 32)\,dx = \left[\frac{0.1x^3}{3} - \frac{4.8x^2}{2} + 32x\right]_0^8$$
$$= 119.47$$

Alternatively, under method 1 or method 2, the statical moment of the area can be obtained by computing the M_x value of the area between line b and the x axis and the M_x value of the area between line a and the x axis. The difference between the two values equals the statical moment of the area that lies between the two lines.

EXAMPLE 10.2 In Fig. 10.3, the curve has the equation $y = 12x - x^2$, and the straight line has the equation $y = 3x$. (Different scales have been used for the x and y axes.) Locate the centroid of the area that lies between the straight line and the curve and that extends from the origin to their point of intersection P.

SOLUTION In computing area and statical moment, we shall follow method 2 of Example 10.1. For this purpose, we divide the region into strips parallel to the y axis, one of which is shown darkened. Again let h denote the height of the strip and dx denote its thickness. Also let x denote the distance from the y axis to the vertical centerline of the strip. Let y_t and y_b denote the distance from the x axis to the top of the strip and to the bottom, respectively.

The steps in the solution are as follows:

1. **Locate the point of intersection P.** Set the two expressions equal to each other to obtain

$$12x - x^2 = 3x \qquad 9x - x^2 = 0 \qquad x(9 - x) = 0$$

The roots of this equation are 0 and 9. Therefore, at P, $x = 9$.

2. **Formulate the expression for the height of the strip**

$$h = y_t - y_b = 12x - x^2 - 3x = 9x - x^2$$

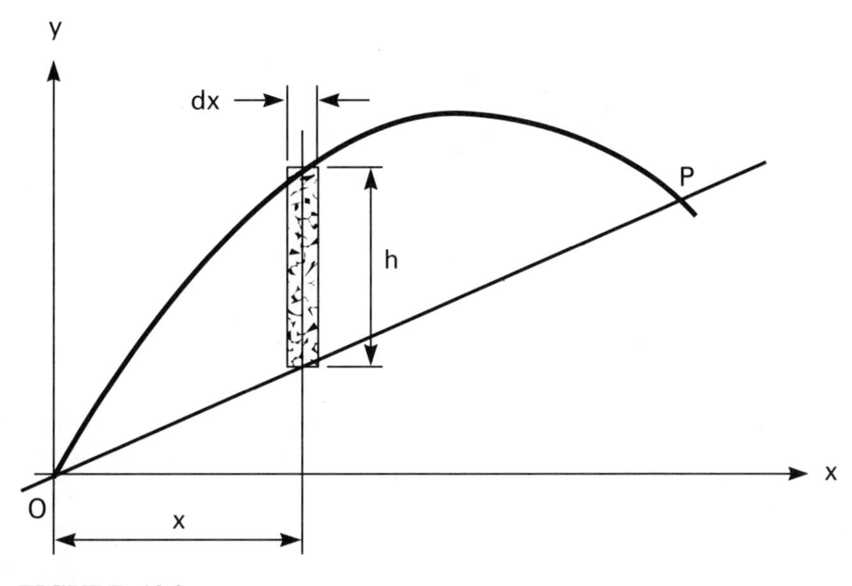

FIGURE 10.3

3. **Compute the amount of area.** The area of the strip is

$$dA = h\,dx = (9x - x^2)\,dx$$

and the total area is

$$A = \int_0^9 (9x - x^2)\,dx = \left[\frac{9x^2}{2} - \frac{x^3}{3}\right]_0^9 = 121.5$$

4. **Compute the statical moment of the area with respect to the x axis.** The statical moment of the strip is

$$dM_x = \left(y_b + \frac{h}{2}\right) dA = \left(3x + \frac{9x - x^2}{2}\right)(9x - x^2)\,dx$$
$$= (67.5x^2 - 12x^3 + 0.5x^4)\,dx$$

Then

$$M_x = \left[\frac{67.5x^3}{3} - \frac{12x^4}{4} + \frac{0.5x^5}{5}\right]_0^9 = 2624.4$$

5. **Compute the statical moment of the area with respect to the y axis.** The statical moment of the strip is

$$dM_y = x\,dA = x(9x - x^2)\,dx = (9x^2 - x^3)\,dx$$

Then

$$M_y = \int_0^9 (9x^2 - x^3)\,dx = \left[\frac{9x^3}{3} - \frac{x^4}{4}\right]_0^9 = 546.75$$

6. **Locate the centroid.** By Eqs. (10.3), the coordinates of the centroid are as follows:

$$y_c = \frac{M_x}{A} = \frac{2624.4}{121.5} = 21.6 \qquad x_c = \frac{M_y}{A} = \frac{546.75}{121.5} = 4.5$$

Comment. It is interesting to observe that the abscissa (x coordinate) of the centroid is half the abscissa of the point of intersection P. This relationship is general for an area of this type, and it can be explained in this manner: If we draw a line u that has the equation $x = 4.5$, we find that two strips that are symmetrically located with respect to u have equal heights. Therefore, their statical moments with respect to u nullify each other, and it follows that u is a centroidal axis.

10.1.3 Moment of Inertia (Second Moment) and Radius of Gyration

Returning to Fig. 10.1, we now form the product $y^2\,dA$ for each element of the area and sum these products. We again make the number of elements n progressively larger by making the elements progressively smaller. The limit of the sum is termed the *moment of inertia* (or *second moment*) of the area with respct to the x axis, and it is denoted by I_x. Then

$$I_x = \int y^2\,dA \qquad (10.4a)$$

Analogously, the moment of inertia of the area with respect to the y axis is

$$I_y = \int x^2 \, dA \tag{10.4b}$$

From the definition, a moment of inertia is restricted to positive values. Its unit is the fourth power of the unit of length.

Let $I_{x,\,c}$ denote the moment of inertia of the area in Fig. 10.1 with respect to its centroidal axis c_x. (This is referred to as a *centroidal* moment of inertia.) Then

$$I_{x,\,c} = \int (y - y_c)^2 \, dA = \int y^2 \, dA - 2y_c \int y \, dA + y_c^2 \int dA$$

$$= I_x - 2y_c M_x + Ay_c^2 = I_x - 2y_c(Ay_c) + Ay_c^2 = I_x - Ay_c^2$$

or

$$I_x = I_{x,\,c} + Ay_c^2 \tag{10.5a}$$

This relationship is known as the *transfer formula.* Similarly, let $I_{y,\,c}$ denote the moment of inertia of the area with respect to its centroidal axis c_y. Then

$$I_y = I_{y,\,c} + Ax_c^2 \tag{10.5b}$$

The moment of inertia of an area is the sum of the moments of inertia of its elements. Therefore, if an area may be considered to be composed of several parts, the moment of inertia of the area is the sum of the moments of inertia of its parts. On the other hand, if part of an area is removed, the moment of inertia of the remaining area is the difference between the moment of inertia of the original area and that of the part removed.

It is advantageous to express moment of inertia in terms of the total area. By analogy with Eqs. (10.2), we may write

$$I_x = Ar_x^2 \qquad I_y = Ar_y^2 \tag{10.6}$$

The distances r_x and r_y that we have introduced for convenience are termed the *radii of gyration* of the area with respect to the x and y axes, respectively. In general, the radius of gyration of the area with respect to an arbitrary axis m is

$$r_m = \sqrt{\frac{I_m}{A}} \tag{10.6a}$$

The radius of gyration of the area with respect to its centroidal axis c_x is denoted by $r_{x,\,c}$.

The radius of gyration is an index of the extent to which the area is dispersed from the indicated axis. However, Eq. (10.5*a*) yields

$$r_x = \sqrt{\frac{I_{x,\,c}}{A} + y_c^2}$$

and it follows that $r_x > y_c$. This condition is explained by the fact that distances are squared in obtaining the radius of gyration. Therefore, the area that lies on the far side of the centroidal axis exerts a greater influence on the radius of gyration than the area that lies on the near side.

The NCEES Handbook presents the properties of standard geometric shapes, such as the rectangle, triangle, and circle. It is helpful to memorize the following properties pertaining to the triangle and the rectangle:

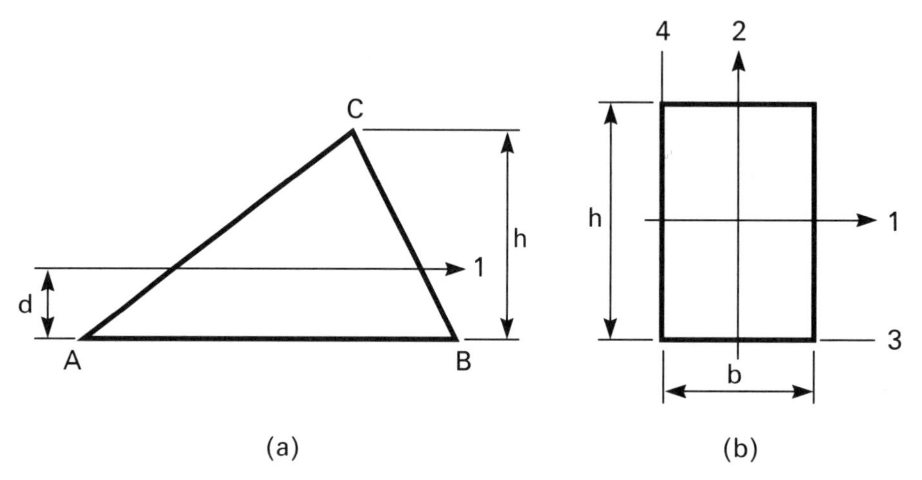

FIGURE 10.4 Significant axes of (*a*) the triangle and (*b*) the rectangle.

Triangle. In Fig. 10.4*a*, axis 1 is the centroidal axis parallel to the base *AB*. Then

$$d = \frac{h}{3} \tag{10.7}$$

Rectangle. In Fig. 10.4*b*, axes 1 and 2 are the centerlines of the rectangle; therefore, they are centroidal axes. The moments of inertia of the area with respect to these axes are as follows:

$$I_1 = \frac{bh^3}{12} \qquad I_2 = \frac{b^3h}{12} \tag{10.8a}$$

The moments of inertia with respect to the sides 3 and 4 are as follows:

$$I_3 = \frac{bh^3}{3} \qquad I_4 = \frac{b^3h}{3} \tag{10.8b}$$

EXAMPLE 10.3 Compute the moment of inertia of the area in Fig. 10.5*a* with respect to its horizontal and vertical centroidal axes.

SOLUTION It is understood that the area is symmetrical about a vertical line, which we have labeled the *y* axis. This is the vertical centroidal axis.

Method 1. Draw a dashed line to separate the stem and the base, thereby decomposing the area into basic shapes (namely, rectangles). Label the stem part 1 and the base part 2. As we perform our calculations, we shall record the values that emerge in Table 10.1.

The steps in the solution are as follows:

1. **Locate the horizontal centroidal axis of the area.** This axis is located by applying Eq. (10.3). Therefore, we must do the following: Compute the area *A*, select a horizontal axis as the *x* axis, and compute the statical moment of the area with respect to this axis. We have selected the base as the *x* axis.

   The area of a part is as follows: part 1, $0.36 \times 1.50 = 0.54\ \text{m}^2$; part 2, $1.20 \times 0.30 = 0.36\ \text{m}^2$. The total area is $0.54 + 0.36 = 0.90\ \text{m}^2$.

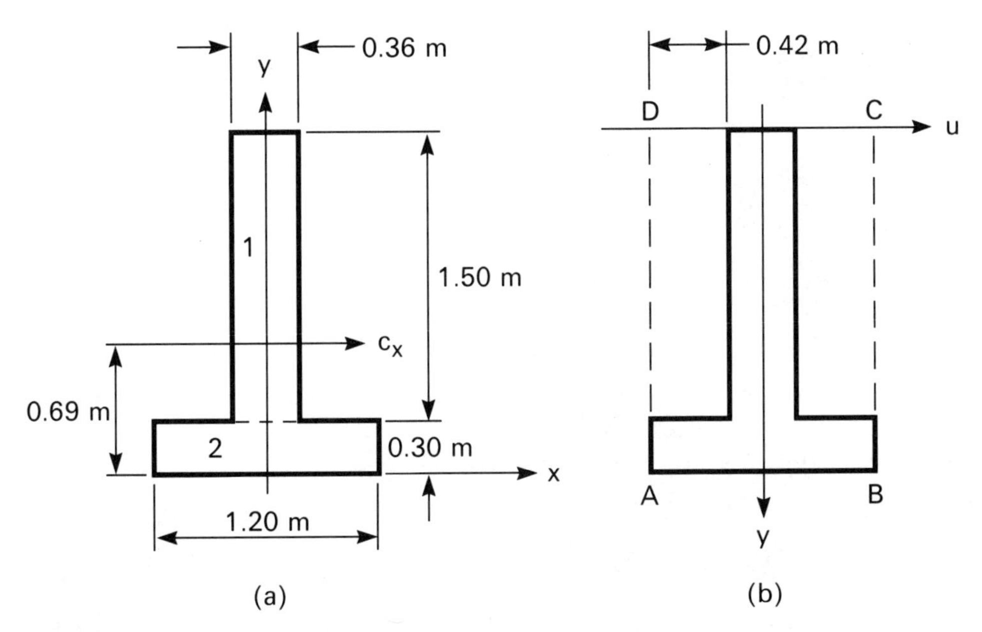

FIGURE 10.5 Given area viewed as (*a*) a combination of areas and (*b*) the difference between two areas.

The distance y_c from the x axis to the horizontal centroidal axis of a part is as follows: part 1, $0.30 + 1.50/2 = 1.05$ m; part 2, $0.30/2 = 0.15$ m.

By Eq. (10.2*a*), the statical moment of a part with respect to the x axis is as follows: part 1, $0.54 \times 1.05 = 0.567$ m^3; part 2, $0.36 \times 0.15 = 0.054$ m^3. The statical moment of the total area is $0.567 + 0.054 = 0.621$ m^3.

For the total area,

$$y_c = \frac{M_x}{A} = \frac{0.621}{0.90} = 0.69 \text{ m}$$

This distance is recorded in Fig. 10.5.

2. **Compute the moment of inertia of the area with respect to its horizontal centroidal axis.** Having located the c_x axis, we can now calculate the moment of inertia with respect to this axis directly, or we can calculate the moment of inertia with respect to the x axis and then apply the transfer formula, Eq. (10.5*a*). We shall follow the latter procedure.

TABLE 10.1 Calculation of I_x

Quantity	Part 1	Part 2	Total area
Area A, m^2	0.54	0.36	0.90
y_c, m	1.05	0.15	0.69
M_x, m^3	0.567	0.054	0.621
I about horizontal centroidal axis of part, m^4	0.10125	0.00270	
Ay_c^2, m^4	0.59535	0.00810	

By Eq. (10.8*a*), the moment of inertia of a part with respect to its horizontal centroidal axis is as follows:

Part 1: $$\frac{(0.36)(1.50)^3}{12} = 0.10125 \text{ m}^4$$

Part 2: $$\frac{(1.20)(0.30)^3}{12} = 0.00270 \text{ m}^4$$

The value of the term Ay_c^2 in Eq. (10.5*a*) is as follows: part 1, $0.54 \times 1.05^2 = 0.59535$ m^4; part 2, $0.36 \times 0.15^2 = 0.00810$ m^4.

By summation, the moment of inertia of the total area with respect to the x axis is

$$I_x = (0.10125 + 0.59535) + (0.00270 + 0.00810) = 0.70740 \text{ m}^4$$

By Eq. (10.5*a*), the moment of inertia of the total area with respect to its horizontal centroidal axis is

$$I_{x,c} = 0.70740 - (0.90)(0.69)^2 = 0.27891 \text{ m}^4$$

3. **Compute the moment of inertia of the area with respect to its vertical centroidal axis.** By Eq. (10.8*a*), we have

$$I_y = \frac{(0.36)^3(1.50)}{12} + \frac{(1.20)^3(0.30)}{12} = 0.04903 \text{ m}^4$$

Comments. The y_c value of the total area is the weighted arithmetic mean of the y_c values of the two parts. The position of the horizontal centroidal axis can be found by an alternative procedure in which we apply the following principle: Let h denote the distance between the horizontal centroidal axes of the two parts. The centroidal axis of the total area divides h into two segments that are inversely proportional to the adjacent areas. In the present instance, $h = 1.05 - 0.15 = 0.90$. Then

$$y_c = 0.15 + 0.90\left(\frac{0.54}{0.90}\right) = 0.69 \text{ m}$$

Method 2. Consider that the given area was obtained by starting with the rectangle $ABCD$ in Fig. 10.5*b* and then removing the two parts at the sides. Let u denote the axis at the top, and take the downward direction as positive. Proceeding as before, we obtain the following results:

$$A = 1.20 \times 1.80 - 2 \times 0.42 \times 1.50 = 2.16 - 1.26 = 0.90 \text{ m}^2$$

$$M_u = 2.16 \times 0.90 - 1.26 \times 0.75 = 0.999 \text{ m}^3$$

The distance from u to the horizontal centroidal axis of the area is

$$y_c = \frac{0.999}{0.9} = 1.11 \text{ m}$$

By Eq. (10.8*b*), the moment of inertia of the area with respect to the *u* axis is

$$I_u = \frac{1.20 \times 1.80^3}{3} - \frac{2 \times 0.42 \times 1.50^3}{3} = 1.3878 \text{ m}^4$$

$$I_{x,c} = 1.3878 - 0.90 \times 1.11^2 = 0.27891 \text{ m}^4$$

The calculation for I_y under method 1 can remain unchanged.

EXAMPLE 10.4 With reference to Fig. 10.6, a circle having a radius of 0.4 m is extracted from the rectangle, at the indicated location. Compute the moment of inertia of the area that remains with respect to its horizontal centroidal axis.

SOLUTION The area that remains can be viewed as the difference between two parts: rectangle *ABCD* and the circle.

The procedure is similar to that in Example 10.3. We place the *x* axis at the base. Refer to the properties of the circle recorded in the NCEES Handbook. The calculations are as follows:

Rectangle

$$A = 1.5 \times 2.0 = 3.0 \text{ m}^2 \qquad M_x = 3.0 \times 1.0 = 3.0 \text{ m}^3$$

By Eq. (10.8b),
$$I_x = \frac{bh^3}{3} = \frac{1.5 \times 2.0^3}{3} = 4.0 \text{ m}^4$$

Circle. Let *r* denote the radius.

$$A = \pi r^2 = \pi(0.4)^2 = 0.5027 \text{ m}^2$$

$$M_x = 0.5027 \times 1.3 = 0.6535 \text{ m}^3$$

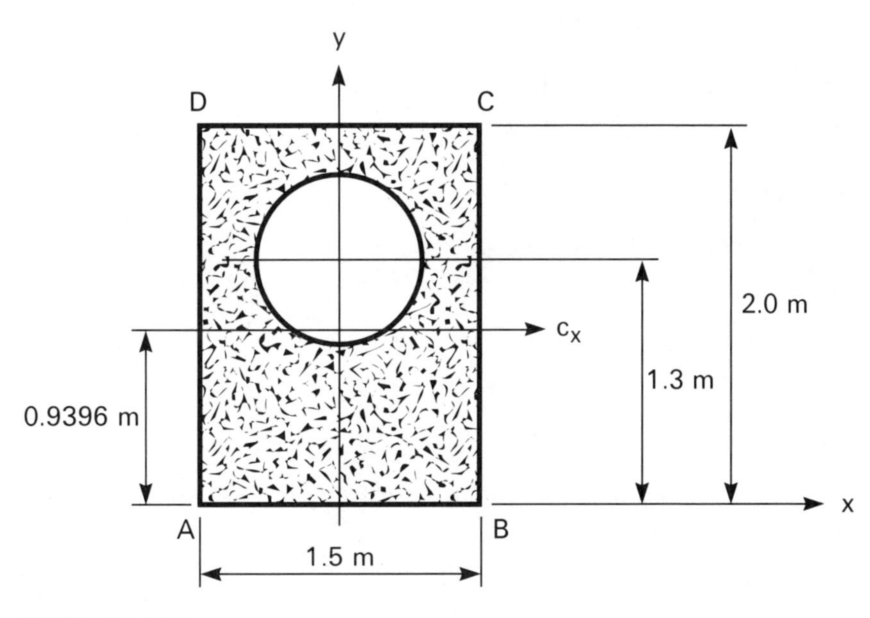

FIGURE 10.6

The moment of inertia of the circle about its centroidal axis is

$$I = \frac{\pi r^4}{4} = \frac{\pi(0.4)^4}{4} = 0.0201 \text{ m}^4$$

The term at the extreme right in Eq. (10.5*a*) has the value

$$Ay_c^2 = (0.5027)(1.3)^2 = 0.8496 \text{ m}^4$$

Then $$I_x = 0.0201 + 0.8496 = 0.8697 \text{ m}^4$$

Net area

$$A = 3.0 - 0.5027 = 2.4973 \text{ m}^2$$

$$M_x = 3.0 - 0.6535 = 2.3465 \text{ m}^3$$

$$y_c = \frac{M_x}{A} = \frac{2.3465}{2.4973} = 0.9396 \text{ m}$$

This distance is recorded in Fig. 10.6.

$$I_x = 4.0 - 0.8697 = 3.1303 \text{ m}^4$$

By Eq. (10.5*a*), the moment of inertia of the net area with respect to its horizontal centroidal axis is

$$I_{x,c} = 3.1303 - 2.4973(0.9396)^2 = 0.9256 \text{ m}^4$$

EXAMPLE 10.5 With reference to Fig. 10.7, the rectangles *abcd* and *efgh* are extracted from the rectangle *ABCD*. The two small rectangles have the same size, and they are symmetrically located in both the horizontal and vertical directions. Compute the moment of inertia of the area that remains with respect to both its horizontal and vertical centroidal axes.

SOLUTION By Theorem 10.1, the centroidal axes c_x and c_y lie at the horizontal and vertical centerlines, respectively, of *ABCD*. The calculations for $I_{x,c}$ are as follows:

Rectangle *ABCD*

$$I_{x,c} = \frac{84 \times 180^3}{12} = 40{,}824 \times 10^3 \text{ mm}^4$$

Rectangle *abcd*

$$A = 24 \times 36 = 864 \text{ mm}^2$$

The moment of inertia of the rectangle with respect to its horizontal centroidal axis is

$$I = \frac{24 \times 36^3}{12} = 93.3 \times 10^3 \text{ mm}^4$$

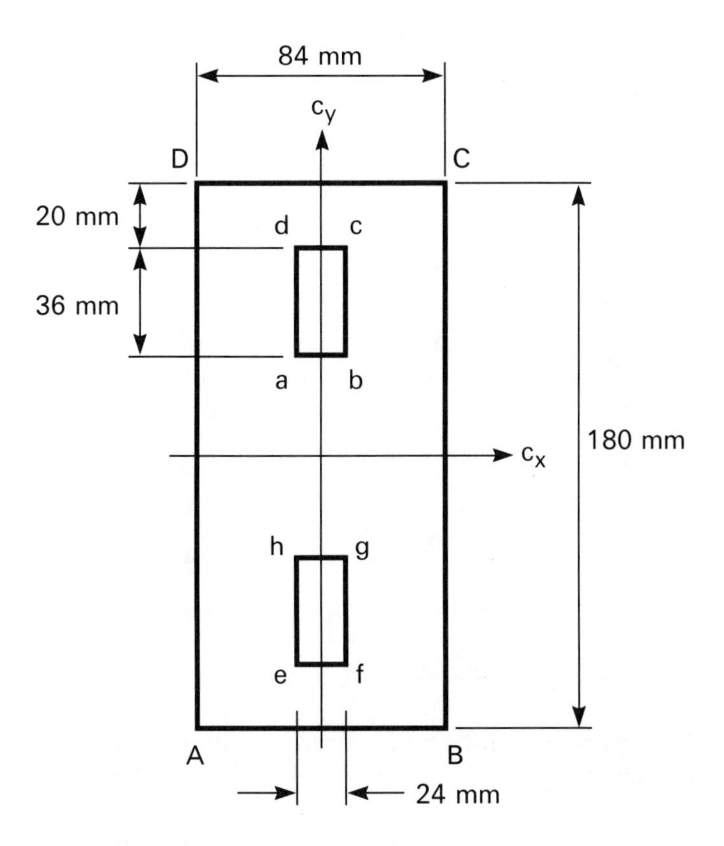

FIGURE 10.7

The distance from the horizontal centroidal axis of *abcd* to that of *ABCD* is

$$y_c = \frac{180}{2} - \left(20 + \frac{36}{2}\right) = 52 \text{ mm}$$

$$Ay_c^2 = 864 \times 52^2 = 2336.3 \times 10^3 \text{ mm}^4$$

$$I_{x,c} = (93.3 + 2336.3)10^3 = 2429.6 \times 10^3 \text{ mm}^4$$

Net area

$$I_{x,c} = (40{,}824 - 2 \times 2429.6)10^3 = 35.96 \times 10^6 \text{ mm}^4$$

The moment of inertia of the net area with respect to the vertical centroidal axis is as follows:

$$I_{y,c} = \frac{84^3 \times 180}{12} - 2\left(\frac{24^3 \times 36}{12}\right) = 8.808 \times 10^6 \text{ mm}^4$$

EXAMPLE 10.6 With reference to Example 10.2, compute the moment of inertia of the area with respect to the y axis.

SOLUTION We compute the moment of inertia of a strip, sum these moments, and allow the number of strips to increase without limit. With reference to the expression for I_y as given by Eq. (10.5*b*), the first term approaches zero as the thickness of the strip approaches zero. Therefore, for the strip,

$$dI_y = x^2\, dA = x^2(9x - x^2)\, dx = (9x^3 - x^4)\, dx$$

For the area,

$$I_y = \int_0^9 (9x^3 - x^4)\, dx = \left[\frac{9x^4}{4} - \frac{x^5}{5}\right]_0^9 = 2952$$

10.1.4 Product of Inertia

Returning to Fig. 10.1, we now form the product $xy\, dA$ for each element, sum these products, and make the number of elements progressively larger. The limit of this sum is called the *product of inertia* of the area with respect to the x and y axes, and it is denoted by I_{xy}. Then

$$I_{xy} = \int xy\, dA \tag{10.9}$$

From the definition, a product of inertia can be positive, negative, or zero. Its unit is the fourth power of the unit of length.

The reasoning that yielded Theorem 10.1 also yields the following principle:

Theorem 10.2. If an area has an axis of symmetry, the product of inertia of the area with respect to this axis and an axis perpendicular thereto is zero.

Let $I_{xc,\, yc}$ denote the product of inertia of an area with respect to its centroidal axes parallel to the x and y axes. Applying the notation shown in Fig. 10.1, we have the following *transfer formula*:

$$I_{xy} = I_{xc,\, yc} + Ax_c y_c \tag{10.10}$$

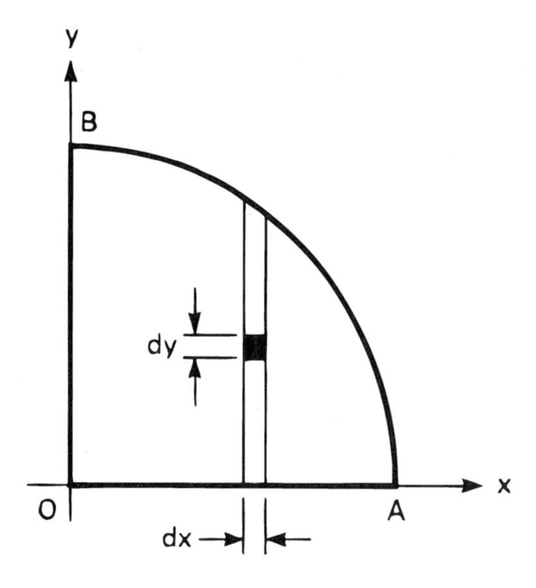

FIGURE 10.8

EXAMPLE 10.7 In Fig. 10.8, OAB is a quadrant of a circle of radius R with its center at the origin. Compute the product of inertia of this area with respect to the x and y axes.

SOLUTION Let dx and dy denote the dimensions of a rectangular element, and let x and y denote the coordinates of its center. The equation of the circle is $x^2 + y^2 = R^2$. Therefore, the x values range from 0 to R, and the y values range from 0 to $\sqrt{R^2 - x^2}$. By double integration, we obtain the following:

$$I_{xy} = \int_0^R \int_0^{\sqrt{(R^2-x^2)}} xy\, dy\, dx$$

$$= \int_0^R \left[x\frac{y^2}{2}\right]_0^{\sqrt{(R^2-x^2)}}$$

$$\int_0^R \frac{x(R^2 - x^2)}{2}\, dx = \int_0^R \left(\frac{R^2x - x^3}{2}\right) dx = \left[\frac{R^2x^2}{4} - \frac{x^4}{8}\right]_0^R = \frac{R^4}{8}$$

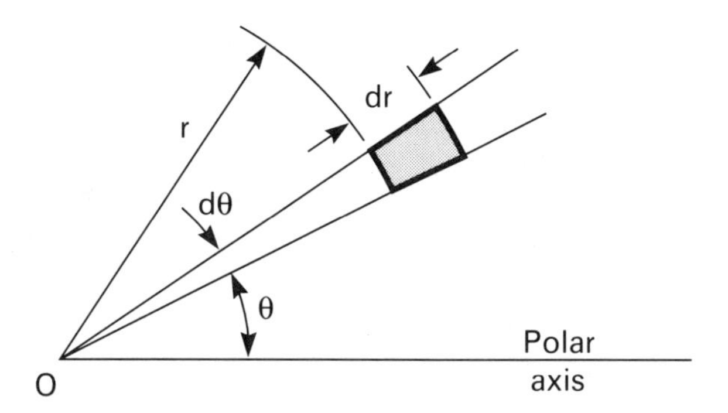

FIGURE 10.9 Use of polar coordinates in computing a property of an area.

The first integration yielded the product of inertia of a strip parallel to the y axis, and the second integration yielded the sum of these quantities.

In many instances, it is more convenient to compute a product of inertia by applying the polar coordinates described in Art. 7.1.6. Refer to Fig. 10.9. We form an element of area bounded by lines through the origin that make angles of θ and $\theta + d\theta$ with the polar axis and by circular arcs having their centers at the origin with radii of r and $r + dr$. Therefore, the length of the element in the radial direction is dr, and its length in the tangential direction may be taken as the length of its inner face, which is $r\,d\theta$. It follows that the area of the element is

$$dA = r\,dr\,d\theta$$

and the product xy may be expressed as

$$xy = (r \cos \theta)(r \sin \theta) = r^2 \sin \theta \cos \theta$$

Then

$$xy\,dA = r^3 \sin \theta \cos \theta\,dr\,d\theta$$

EXAMPLE 10.8 Solve Example 10.7 by applying polar coordinates.

SOLUTION

$$I_{xy} = \int_0^{\pi/2} \int_0^R r^3 \sin \theta \cos \theta\,dr\,d\theta$$

$$= \frac{R^4}{4} \int_0^{\pi/2} \sin \theta \cos \theta\,d\theta = \frac{R^4}{4} \left[\frac{\sin^2 \theta}{2} \right]_0^{\pi/2} = \frac{R^4}{8}$$

10.1.5 Effect of Rotation of Axes

Assume the following: We are given an area and the positions of the x and y axes; we know I_x, I_y, and I_{xy}; we also require the values of moment of inertia and product of inertia with respect to other axes through the origin. To obtain these values, consider that the x and y axes are rotated about the origin in a counterclockwise direction through an angle θ. Let x' and y' denote the new positions of the axes.

By formulating the expressions for $I_{x'}$ and $I_{x'y'}$ and allowing θ to vary, we deduce the following:

1. There is a value of θ for which $I_{x'}$ is maximum and one for which $I_{x'}$ is minimum. The corresponding axes are referred to collectively as the *principal axes* through the origin. In particular, the axis for which moment of inertia is maximum is called the *major axis*, and that for which moment of inertia is minimum is called the *minor axis*.
2. The principal axes are mutually perpendicular.
3. The product of inertia with respect to the principal axes is zero.
4. Similarly, there is a value of θ for which $I_{x'y'}$ is maximum and one for which $I_{x'y'}$ is minimum. These axes are also mutually perpendicular, and they lie midway between the principal axes.

As an illustration, refer to the rectangle in Fig. 10.4*b*. The axes of symmetry, 1 and 2, are the principal axes through the center. If $h > b$, axis 1 is the major axis and axis 2 is the minor axis. The moment of inertia of the rectangle with respect to any other centroidal axis has some value intermediate between I_1 and I_2.

To illustrate the significance of the foregoing principles, we shall consider a problem that arises in structural engineering: determining the capacity of a column. Assume that a column that lacks bracing is subjected to a gradually increasing load until it buckles. When the column buckles, each cross section rotates about its minor centroidal axis. Therefore, it is necessary to identify this axis and to compute the moment of inertia of the section with respect to this axis.

10.1.6 Polar Moment of Inertia

Returning to Fig. 10.1, let z denote an axis through the origin that is perpendicular to the xy plane, and let ρ denote the distance from the center of an element to axis z. We now form the product $\rho^2\,dA$, sum these products, and make the number of elements progressively larger. The limit of this sum is called the *polar moment of inertia* of the area with respect to the z axis, and it is denoted by J_z. Then

$$J_z = \int \rho^2\,dA \qquad (10.11)$$

From the definition, a polar moment of inertia is restricted to positive values, and its unit is the fourth power of the unit of length. Since $\rho^2 = x^2 + y^2$, we have

$$J_z = I_x + I_y \qquad (10.12)$$

Consider again that the x and y axes are rotated about the origin to the positions x' and y'. Since J_z remains constant, it follows that the sum $I_{x'} + I_{y'}$ is constant for all values of the angle of rotation.

Now let c_z denote the axis that is parallel to z and passes through the centroid of the area, $J_{z,c}$ denote the polar moment of inertia of the area with respect to c_z, and ρ_c denote the distance from the centroid of the area to the z axis. The *transfer formula* for polar moment of inertia is

$$J_z = J_{z,c} + A\rho_c^2 \qquad (10.13)$$

The radius of gyration of the area with respect to the z axis is

$$r_z = \sqrt{\frac{J_z}{A}} \tag{10.14}$$

EXAMPLE 10.9 A circle of radius 3 has its center at $C(7,5)$. Compute the radius of gyration of the area with respect to an axis that is perpendicular to the plane of the area and contains the point $P(-4,9)$.

SOLUTION We can displace the origin to P, and the coordinates of the center become $x = 7 - (-4) = 11$ and $y = 5 - 9 = -4$. Let r denote the radius, and refer to the NCEES Handbook for the properties of a circle.

$$\rho_c^2 = 11^2 + (-4)^2 = 137 \qquad A = \pi r^2 = \pi \times 3^2 = 28.27$$

$$J_{z,c} = \frac{\pi r^4}{2} = \frac{\pi \times 3^4}{2} = 127.23$$

$$J_z = 127.23 + 28.27 \times 137 = 4000$$

$$r_z = \sqrt{\frac{4000}{28.27}} = 11.90$$

10.2 PROPERTIES OF ARCS, VOLUMES, AND MASSES

10.2.1 Properties of an Arc

We have defined the properties of an area, and an arc of a curve has analogous properties. Let s denote the length of the arc. Consider that the arc is divided into elements of length ds. The statical moments of the arc are as follows:

$$M_x = \int y\,ds \qquad M_y = \int x\,ds$$

By applying Eq. (8.6), we obtain the following:

$$M_x = \int y\sqrt{1 + \left(\frac{dx}{dy}\right)^2}\,dy \tag{10.15a}$$

$$M_y = \int x\sqrt{1 + \left(\frac{dy}{dx}\right)^2}\,dx \tag{10.15b}$$

Again let c_x and c_y denote axes that are parallel to the x and y axes, respectively, and that have this property: The statical moment of the arc with respect to these axes is zero. Axes c_x and c_y are centroidal axes of the arc. With due regard for algebraic signs, let y_c denote the distance between the c_x and x axes, and let x_c

denote the distance between the c_y and y axes. Then

$$x_c = \frac{M_y}{s} \qquad y_c = \frac{M_x}{s} \tag{10.16}$$

10.2.2 Properties of a Volume

The volume of a solid body has properties analogous to those of an area. However, for a solid body, the statical moment is taken with respect to a *plane*, and the moment of inertia may be taken with respect to either a plane or a line.

Consider that a body of volume V is divided into elements of volume dV. The statical moment and moment of inertia of the entire volume with respect to the xy plane are as follows:

$$M_{xy} = \int z\,dV \qquad I_{xy} = \int z^2\,dV$$

Let c_{xy} denote a plane that is parallel to the xy plane and that has this property: The statical moment of the volume with respect to plane c_{xy} is zero. The c_{xy} is a *centroidal plane*, and its distance z_{xy} from the xy plane is

$$z_{xy} = \frac{M_{xy}}{V} \tag{10.17}$$

Analogous equations apply with respect to the other coordinate planes. The point of intersection of the three centroidal planes parallel to the coordinate planes is the centroid of the volume.

Let r_x denote the distance from the center of an element to the x axis. The moment of inertia of the volume with respect to the x axis is

$$I_x = \int r_x^2\,dV = \int (y^2 + z^2)\,dV = I_{xz} + I_{xy}$$

EXAMPLE 10.10 Locate the centroid of a solid in the form of a hemisphere of radius R.

SOLUTION Place the coordinate axes in the position shown in Fig. 10.10, where the origin is at the center of the corresponding sphere. Since they are planes of symmetry, the xy and xz planes are centroidal planes. Therefore, $y_{xz} = z_{xy} = 0$, and it is simply necessary to find x_{yz}. To do this, we must find M_{yz}.

The problem can be solved by setting $dM_{yz} = x\,dx\,dy\,dz$ and then applying triple integration. However, a more direct approach consists of evaluating the derivative dM_{yz}/dx. We divide the body into strips parallel to the yz plane, of width dx, and consider the radius r of a strip to be constant, as shown in Fig. 10.10. Let A denote the area of the face of a strip. The line of intersection of the solid and the xz plane is a semicircle of radius R, and its equation is $x^2 + z^2 = R^2$. For a particular strip, we have the following:

$$dM_{yz} = x\,dV = xA\,dx = x(\pi z^2)\,dx = \pi x(R^2 - x^2)\,dx$$

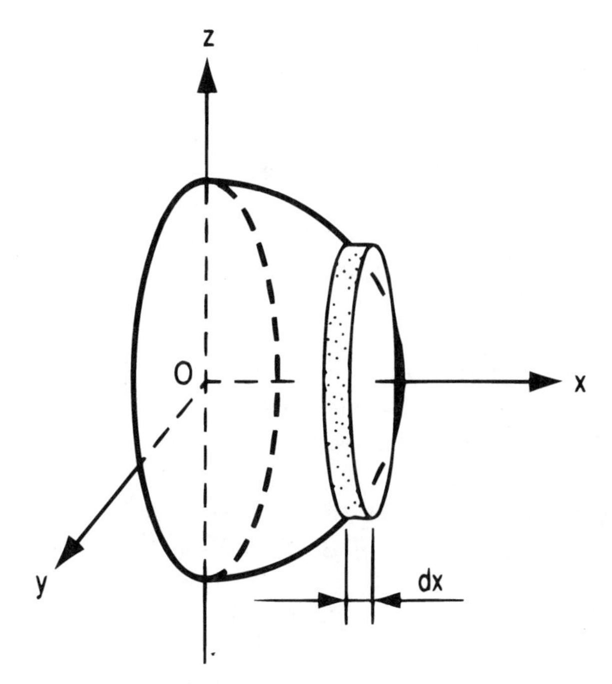

FIGURE 10.10

$$M_{yz} = \pi \int_0^R (R^2x - x^3)\, dx = \pi \left[\frac{R^2x^2}{2} - \frac{x^4}{4}\right]_0^R = \frac{\pi R^4}{4}$$

The NCEES Handbook presents the expression for the mass of a sphere (at the end of the "Dynamics" section). If we divide this expression by the density of the material, we obtain $4\pi R^3/3$ as the volume of the sphere. Therefore, the volume of a hemisphere is $2\pi R^3/3$. Then

$$x_{yz} = \frac{M_{yz}}{V} = \frac{\pi R^4/4}{2\pi R^3/3} = \frac{3}{8} R$$

10.2.3 Properties of a Mass

The properties of a mass are analogous to those of a volume. However, the centroid of the body is termed its *mass center* or *center of gravity.*

Let ρ denote the density of the material (i.e., the mass of a unit volume), m denote the mass of the body, and dm denote the mass of an element of the body. The statical moment of the mass with respect to the xy plane is

$$M_{xy} = \int z\, dm = \int z\rho\, dV$$

If a body is homogeneous (i.e., made of a single material), its mass center coincides with the centroid of its volume. When a moment (or torque) acts on a body, the body rotates, and its angular acceleration is inversely proportional to its moment of inertia with respect to the axis of rotation.

Let I_c denote the moment of inertia of a body with respect to a centroidal axis, let u denote an axis parallel thereto, and let d denote the distance between the two axes. The moment of inertia with respect to u is

$$I_u = I_c + md^2 \tag{10.18}$$

Although the NCEES Handbook presents the properties of a body in the form of a right circular cylinder, we shall nevertheless compute a moment of inertia of this body to illustrate the procedure. Moreover, an examination problem may require simply that the examinee formulate the expression for a given quantity in the form of integrals.

EXAMPLE 10.11 A homogeneous right circular cylinder has a length h and a radius R. Compute the moment of inertia of this body with respect to its longitudinal axis, expressing the result in terms of the mass of the cylinder.

SOLUTION This problem lends itself to solution by polar coordinates. Refer to the discussion in Art 10.1.4. We form a strip of volume that extends across the entire length of the cylinder and that has the section shown in Fig. 10.9. Let r denote the distance from this element to the longitudinal axis. The strip has the following volume, mass, and moment of inertia with respect to the longitudinal axis:

$$dV = hr\,dr\,d\theta \qquad dm = \rho hr\,dr\,d\theta$$

$$dI = r^2\,dm = \rho hr^3\,dr\,d\theta$$

The total mass is

$$m = \rho h \int_0^{2\pi} \int_0^R r\,dr\,d\theta = \rho h \frac{R^2}{2} \int_0^{2\pi} d\theta = \pi \rho h R^2$$

The total moment of inertia is

$$I = \rho h \int_0^{2\pi} \int_0^R r^3\,dr\,d\theta = \rho h \frac{R^4}{4} \int_0^{2\pi} d\theta = \pi \rho h \frac{R^4}{2}$$

Dividing I by m, we obtain

$$I = \frac{mR^2}{2}$$

10.2.4 Theorems of Pappus

The following principles are helpful in calculating surface areas and volumes that result from revoluton:

Theorem 10.3. If the arc of a plane curve is revolved about an axis that lies in the same plane but does not intersect the arc, the area of the surface generated is equal to the product of the length of the arc and the distance traversed by the centroid of the arc.

Theorem 10.4. If a plane region is revolved about an axis that lies in that plane but does not lie within this region, the volume generated is equal to the product of the area of the region and the distance traversed by the centroid of the region.

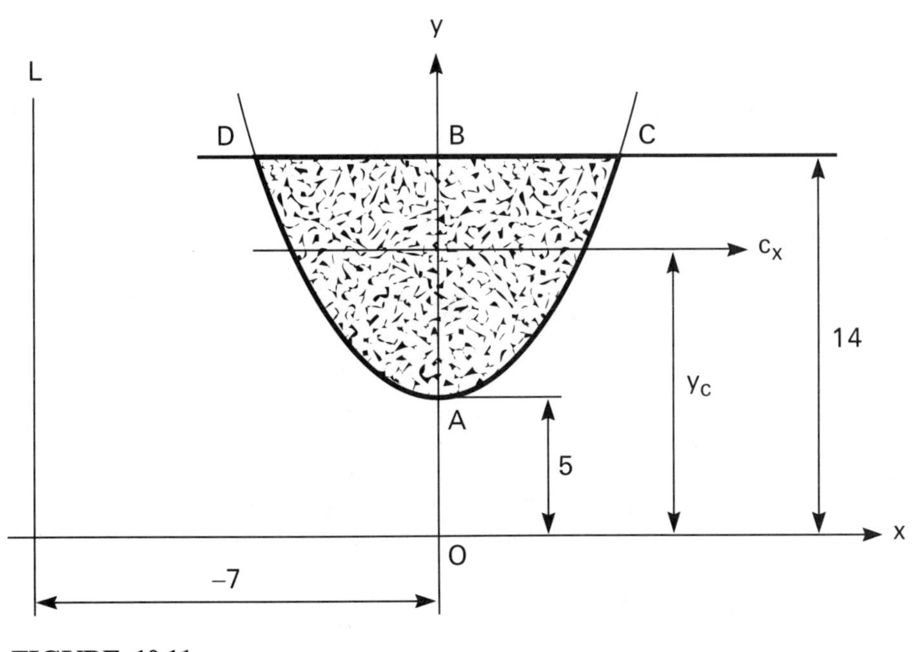

FIGURE 10.11

EXAMPLE 10.12 The curve in Fig. 10.11 has the equation $y = x^2 + 5$. (Different scales have been applied to the x and y axes.) The area bounded by this curve and the line $y = 14$ is revolved about the x axis through an angle of 2π. What volume is generated?

SOLUTION The curve is a parabola, and it generates a hollow body when it is revolved about the x axis. When $x = 0$, $y = 5$. When $y = 14$, $x = \pm 3$. Then $AB = 14 - 5 = 9$ and $BC = 3$.

Refer to the NCEES Handbook for the properties of the parabola. In the present instance, $a = AB = 9$ and $b = BC = 3$. The area is

$$A = \frac{4ab}{3} = \frac{4 \times 9 \times 3}{3} = 36$$

$$y_c = \frac{3a}{5} + 5 = \frac{3 \times 9}{5} + 5 = 10.4$$

As the area is revolved, the centroid of the area describes a circle of radius y_c, and the perimeter of the circle is $2\pi y_c$. By Theorem 10.4, the volume generated is

$$V = 2\pi y_c A = 2\pi \times 10.4 \times 36 = 748.8\pi$$

EXAMPLE 10.13 With reference to Example 10.12, what volume is generated if the area is revolved about line L, which has the equation $x = -7$, through an angle of 2π?

SOLUTION Line L lies outside the given area, and the distance d from the centroid of the area to L is 7. Then

$$V = 2\pi dA = 2\pi \times 7 \times 36 = 504\pi$$

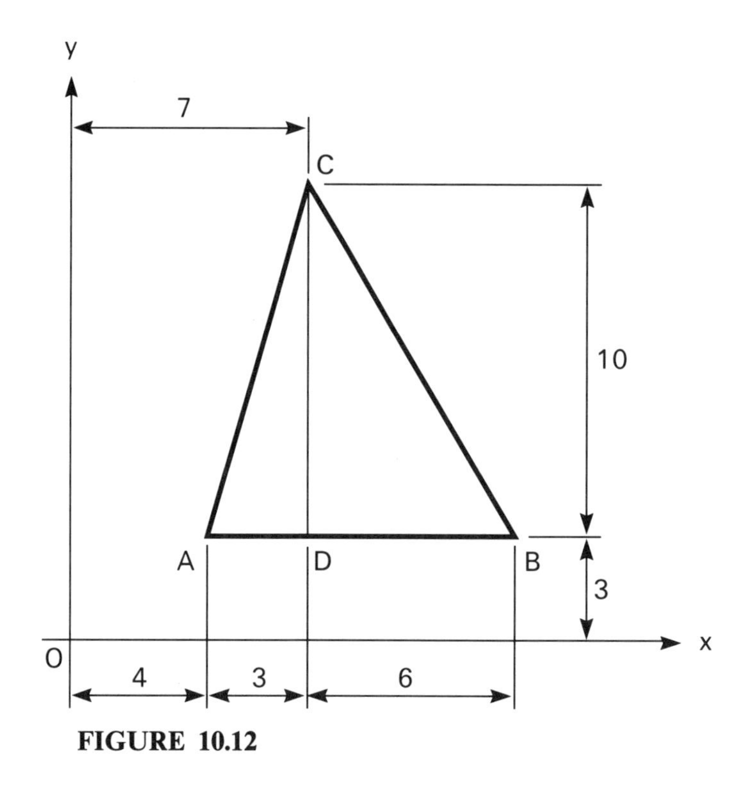

FIGURE 10.12

EXAMPLE 10.14 A triangle has its vertices at $A(4,3)$, $B(13,3)$, and $C(7,13)$. The triangle is revolved about the line $x = 7$ through an angle of π. What volume is generated?

SOLUTION Refer to Fig. 10.12, where we have recorded the relevant dimensions. Draw DC parallel to the y axis, thereby dividing triangle ABC into the triangles ADC and BDC. Line DC is the axis of revolution. Let the subscripts 1, 2, and T refer to triangles ADC, BDC, and ABC, respectively.

Method 1. Each of the smaller triangles generates half of a right circular cone. From the NCEES Handbook, the volume of a right circular cone is $\pi r^2 h/3$. Therefore, the volume generated is

$$V_T = \frac{\pi \times 10}{6}(3^2 + 6^2) = 75\pi$$

Method 2. We shall apply Theorem 10.4. The distance from the centroid of a triangle to line DC is 1 for ADC and 2 for BDC. The areas are as follows:

$$A_1 = (1/2)3 \times 10 = 15 \qquad A_2 = 30$$

Each centroid traverses a semicircular arc, and the volumes generated are as follows:

$$V_1 = \pi \times 1 \times 15 = 15\pi \qquad V_2 = \pi \times 2 \times 30 = 60\pi$$

Then

$$V_T = \pi(15 + 60) = 75\pi$$

EXAMPLE 10.15 Solve Example 9.16 by applying Theorem 10.4.

SOLUTION The curve in Fig. 9.7 is a half-parabola. Refer to the NCEES Handbook.

$$a = OA = X \qquad b = AB = X^{1/2}$$

$$A = \frac{2ab}{3} = \frac{2X^{3/2}}{3} \qquad x_c = \frac{3a}{5} = \frac{3X}{5}$$

The volume generated is

$$V = 2\pi x_c A = 2\pi \frac{3X}{5} \frac{2X^{3/2}}{3} = \frac{4\pi X^{5/2}}{5}$$

EXAMPLE 10.16 When a circle is revolved about an axis that lies in the same plane but outside the circle, the solid thus generated is termed a *torus*. If r is the radius of the circle and R is the distance from the center of the circle to the axis of revolution, what is the volume of the torus?

SOLUTION The center of the revolving circle traverses a circle having a radius of R.

$$A = \pi r^2 \qquad V = 2\pi R A = 2\pi^2 R r^2$$

CHAPTER 11

STATISTICS AND PROBABILITY OF CONTINUOUS VARIABLES

In Chaps. 4 and 5, we explored the subjects of statistics and probability, but we restricted our study to discrete variables. We shall now apply the principles and techniques of calculus to extend our study of these subjects to continuous variables.

11.1 STATISTICS

11.1.1 Frequency Distributions and Frequency Curves

Again let X denote a quantity that can assume different values on different occasions. In Art. 4.1, we defined the terms *relative frequency, cumulative frequency,* and *frequency distribution* when X is discrete, and in Fig. 4.2 we exhibited the frequency distribution of a variable in the form of a *histogram* when the values assumed by X are grouped in classes. In a histogram, the area of each rectangle equals the frequency of the corresponding class.

Now assume that X is a continuous variable and that we have compiled an infinite set of values of X. To exhibit the frequency distribution of X, we may proceed in this manner: Divide the range of X values into classes and construct a histogram in which the area of each rectangle equals the *relative frequency* of the corresponding class. Now make the width of each class progressively smaller. In the limit, the histogram approaches a smooth curve, and this is referred to as the *frequency curve* of X.

The basic characteristic of a frequency curve is that the relative frequency corresponding to a given range of values of X equals the area bounded by the curve, the horizontal axis, and vertical lines erected at the boundaries of the range. For example, assume that X can range from 2.0 to 16.0 and has the frequency curve in Fig. 11.1. The relative frequency of values of X ranging from 5.0 to 8.0 equals the shaded area in the diagram, and the total area under the curve is 1.

The ordinate of a frequency curve is denoted by $f(X)$, and it is called the *frequency-density function.* Let $F(a,b)$ denote the relative frequency of values of X ranging from a to b. Then

$$F(a,b) = \int_a^b f(X)\, dX \qquad (11.1)$$

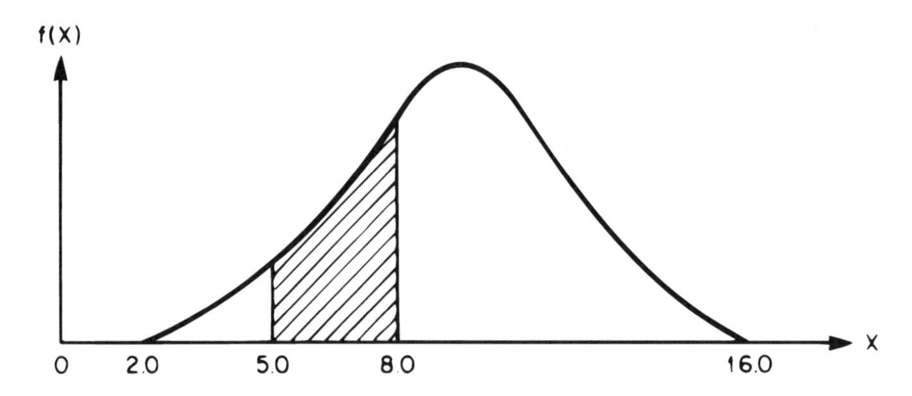

FIGURE 11.1 Frequency curve.

Now let $F(a)$ denote the relative frequency of all values of X less than or equal to a. Thus, $F(a)$ is a *cumulative* relative frequency, and it is termed the *frequency-distribution function.* Then

$$F(a) = \int_{-\infty}^{a} f(X)\,dX \tag{11.2}$$

Figure 11.2 illustrates three basic types of frequency curves. The bell-shaped curve in Fig. 11.2*a* is symmetrical about a vertical axis. The curves in Figs. 11.2*b* and *c* are unsymmetrical, and they are accordingly described as *skewed.* If the longer tail of the curve lies to the right of the summit, as in Fig. 11.2*b*, the curve is *skewed to the right*; if the longer tail lies to the left of the summit, as in Fig. 11.2*c*, the curve is *skewed to the left.*

11.1.2 Arithmetic Mean

In Art. 4.5, we defined the arithmetic mean $\bar{X}$ of a discrete variable X, and the value of $\bar{X}$ is given by Eq. (4.1). We shall now develop the counterpart of this equation for a continuous variable.

Consider that X is a function of a continuous variable U, that U can range from a to b, and that the relationship between X and U is as shown in Fig. 11.3. If we compute the area under this curve and then divide this area by $b - a$, we obtain the *mean ordinate* of the curve, and this is the arithmetic mean of X. Then

$$\bar{X} = \frac{\int_a^b X\,dU}{b-a} \tag{11.3}$$

EXAMPLE 11.1 A variable X has the equation $X = A \sin \theta$, where A is a constant. Find the arithmetic mean of X when θ varies from 0 to π. Devise an approximate check.

SOLUTION The integral in Eq. (11.3) becomes

$$A \int_0^{\pi} \sin \theta \, d\theta = A[-\cos \theta]_0^{\pi} = A[-(-1-1)] = 2A$$

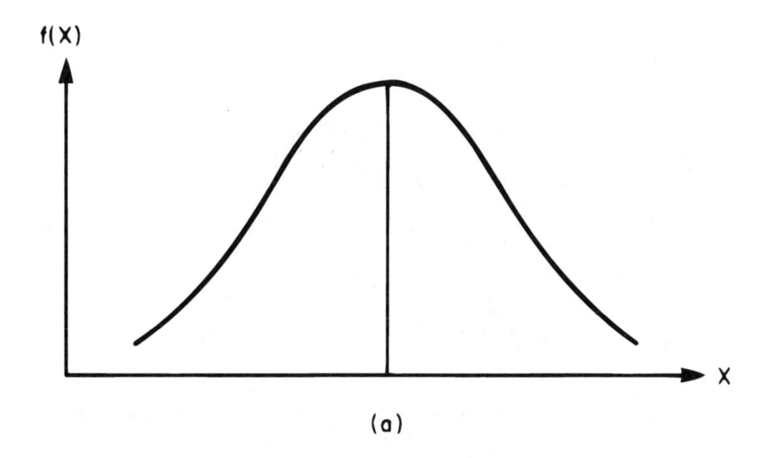

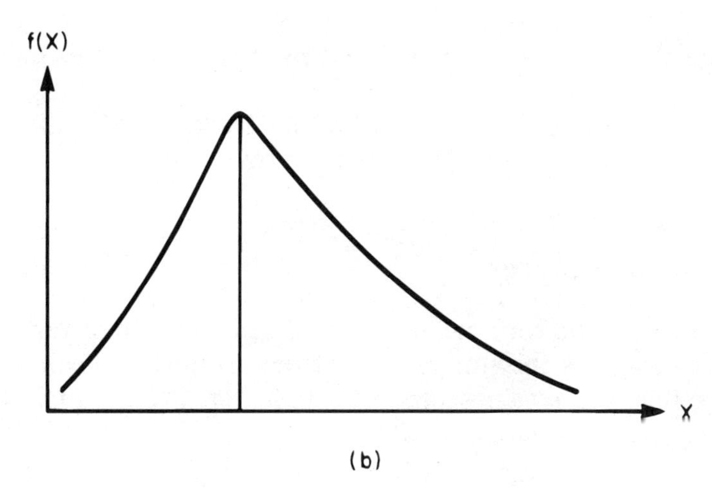

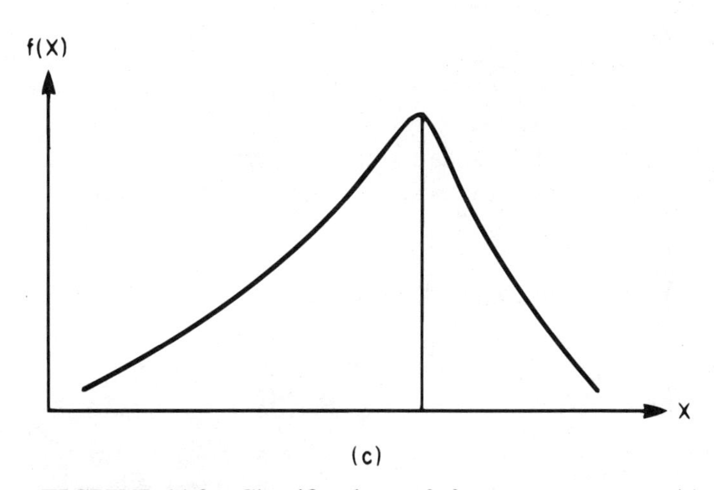

FIGURE 11.2 Classification of frequency curves. (*a*) Symmetrical curve; (*b*) curve skewed to the right; (*c*) curve skewed to the left.

Then
$$\bar{X} = \frac{2A}{\pi} = 0.6366A$$

As an approximate check, assume that X varied *linearly* from 0 to A and then back to 0. The arithmetic mean of X would be $0.5A$. From the shape of the sine

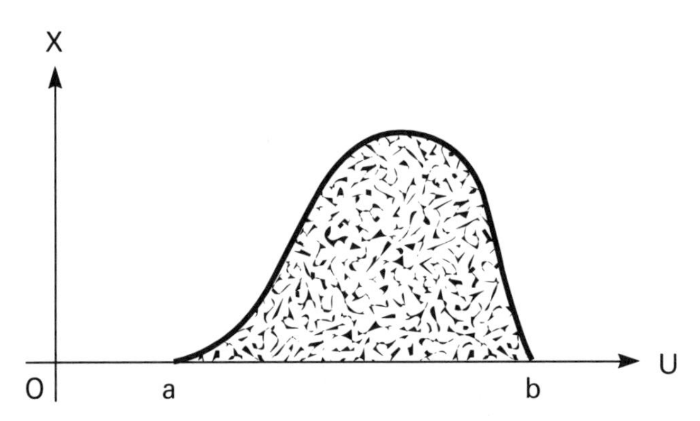

FIGURE 11.3 Establishing the arithmetic mean of a continuous variable.

curve, we deduce that the true arithmetic mean must exceed this value by a small amount, and it does.

The range of θ could have been taken as 0 to $\pi/2$ without affecting $\bar{X}$ because the X values repeat themselves beyond that point (in reverse order).

11.1.3 Root Mean Square

In Art. 4.10, we defined the root mean square X_{rms} of a discrete variable X, and we stated that this quantity is the square root of the arithmetic mean of the values of X^2. Therefore, to find X_{rms}, we proceed in this manner: Modify Fig. 11.3 by plotting X^2 on the vertical axis, compute the total area under the curve, and divide by $b - a$. Then

$$X_{rms} = \sqrt{\frac{\int_a^b X^2\, dU}{b - a}} \tag{11.4}$$

EXAMPLE 11.2 Find the root mean square of the variable in Example 11.1 when θ varies from 0 to 2π. Express the result in terms of the maximum value assumed by X.

SOLUTION The integral in Eq. (11.4) becomes

$$A^2 \int_0^{2\pi} \sin^2 \theta\, d\theta = A^2 \left[\frac{\theta}{2} - \frac{\sin 2\theta}{4}\right]_0^{2\pi} = A^2 \left(\frac{2\pi}{2}\right) = A^2 \pi$$

Then

$$X_{rms} = \sqrt{\frac{A^2 \pi}{2\pi}} = \frac{A}{\sqrt{2}} = 0.7071A$$

Let X_{max} denote the maximum value of X. Then $X_{max} = A$, and the foregoing result becomes

$$X_{rms} = \frac{X_{max}}{\sqrt{2}} = 0.7071X_{max}$$

The range of θ could have been taken as 0 to π because the values of X^2 for this interval repeat themselves cyclically. The result we have obtained is applied in electrical engineering, where a sinusoidally varying voltage or current is expressed in terms of its rms value.

11.2 PROBABILITY

11.2.1 Probability Distributions and Probability Curves

In Art. 5.6, we defined the *probability distribution* of a discrete random variable. This expresses the probability that the variable will assume a *particular value* on a given occasion. Similarly, the probability distribution of a continuous random variable expresses the probability that the variable will assume a value that lies within a *particular interval* on a given occasion. We shall again let X denote the random variable.

The probability distribution of a continuous variable can be exhibited visually by constructing a *probability curve*, in which values of X are plotted on the horizontal axis. The curve has this property: The area bounded by the curve, the X axis, and vertical lines erected at $X = a$ and at $X = b$ equals the probability that $a \leq X \leq b$. For example, with reference to Fig. 11.4, the shaded area is the probability that $1.40 \leq X \leq 2.05$.

Let $f(X)$ denote the ordinate of the probability curve. Then

$$P(a \leq X \leq b) = \int_a^b f(X)\, dX = \text{area under probability curve between } a \text{ and } b$$

The ordinate $f(X)$ is referred to as the *probability-density function.* Now let $F(a)$ denote the probability that $X \leq a$. Then

$$F(a) = P(X \leq a) = \int_{-\infty}^{a} f(X)\, dX = \text{area under probability curve to left of } a$$

The quantity $F(a)$ is termed the *probability-distribution function.*

11.2.2 Characteristics of a Probability Distribution

Since it is a certainty that X will assume one of its possible values and the probability of certainty is 1, it follows that the total area under a probability curve is 1.

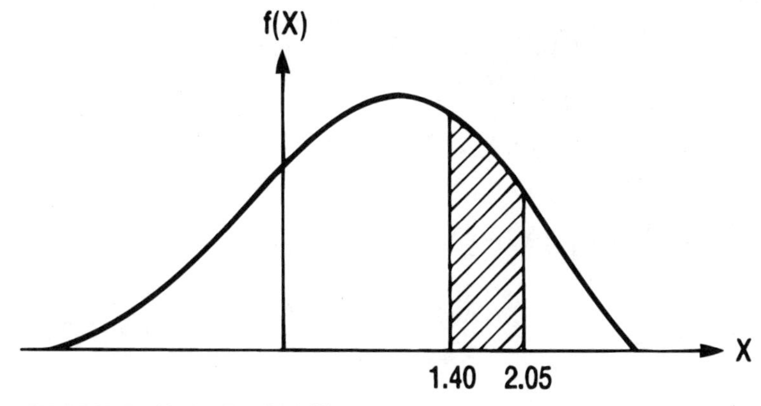

FIGURE 11.4 Probability curve.

Thus,

$$\int_{-\infty}^{+\infty} f(X)\,dX = 1 \qquad (11.5)$$

In Art. 5.7, we defined the arithmetic mean μ and standard deviation σ of a discrete probability distribution. The quantity σ^2 is known as the *variance*. To transform Eq. (5.2) to its counterpart for a continuous variable, we follow this formula: Replace $P(X)$, which is the probability of a particular value of X, with $f(X)\,dX$, which, when integrated, is the probability of a particular interval of X values. The result is

$$\mu = \int_{-\infty}^{+\infty} [f(X)]X\,dX \qquad (11.6)$$

This formula also transforms Eq. (5.3) to the following expression for the variance:

$$\sigma^2 = \int_{-\infty}^{+\infty} [f(X)](X-\mu)^2\,dX \qquad (11.7)$$

EXAMPLE 11.3 A continuous random variable X has the following probability-density function:

$$f(X) = 0.024X^2 \qquad \text{if } 0 < X \le 5$$

$$f(X) = 0 \qquad \text{elsewhere}$$

a. Prove that the equation for $f(X)$ is valid.

b. Compute the arithmetic mean and standard deviation of this distribution.

c. Compute the probability that X will assume a value lying between 2 and 4 on a given occasion.

SOLUTION

Part a

$$\int_0^5 f(X)\,dX = \frac{(0.024)5^3}{3} = 1$$

Since Eq. (11.5) is satisfied, the equation for $f(X)$ is valid.

Part b. By Eqs. (11.6) and (11.7),

$$\mu = 0.024\int_0^5 X^3\,dX = \frac{(0.024)5^4}{4} = 3.75$$

$$\sigma^2 = 0.024\int_0^5 X^2(X-3.75)^2\,dX$$

$$= 0.024\int_0^5 (X^4 - 7.50X^3 + 14.0625X^2)\,dX$$

$$= 0.024\left(\frac{5^5}{5} - \frac{7.50\times 5^4}{4} + \frac{14.0625\times 5^3}{3}\right) = 0.9375$$

$$\sigma = \sqrt{0.9375} = 0.968$$

Part c

$$P(2 \le X \le 4) = \int_2^4 f(X)\, dX = \frac{(0.024)(4^3 - 2^3)}{3} = 0.448$$

11.2.3 Normal Distribution

A continuous random variable is said to have a *normal* or *gaussian probability distribution* if the range of its possible values is infinite and its probability-density function has this form:

$$f(X) = \frac{1}{b\sqrt{2\pi}} e^{-K}$$

where

$$K = \frac{(X-a)^2}{2b^2} \tag{11.8}$$

In this equation, e is the quantity defined in Art. 1.20, and a and b are constants.

Equation (11.8) was first formulated by DeMoivre in solving a problem in gambling, but it was later discovered that this equation also describes a vast number of random variables that appear in natural phenomena.

Figure 11.5 is the normal probability curve for assumed values of a and b. This bell-shaped curve has a summit at $X = a$, and it is symmetrical about the vertical line through the summit. Thus, the constant a in Eq. (11.8) is the arithmetic mean of the normal probability distribution. It can also be demonstrated that the constant b is the standard deviation. Expressed symbolically,

$$\mu = a \qquad \sigma = b$$

It is convenient to establish a variable x that is associated with X in the following manner:

$$x = \frac{X - \mu}{\sigma} \tag{11.9}$$

Thus, when $X = \mu$, $x = 0$. The probability distribution of x is termed the *unit* normal distribution.

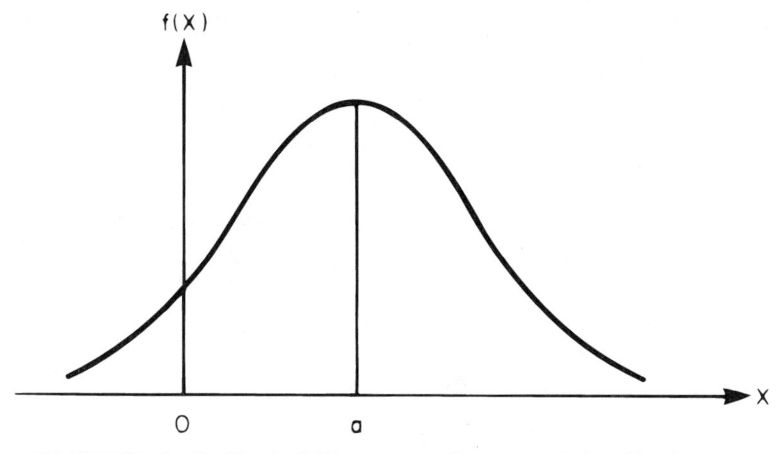

FIGURE 11.5 Probability curve of a normal distribution.

The NCEES Handbook contains a table that gives the area under the unit normal curve corresponding to a given value of x. As indicated by the shading, the column labeled $F(x)$ gives the area that lies to the *left* of the given x value. Therefore, it presents the probability that x is equal to or less than the specified value. The column labeled $R(x)$ gives the area that lies to the *right* of the given x value. Therefore, it presents the probability that x is equal to or greater than the specified value. Since the curve is symmetrical, it follows that $P(x \leq 0) = P(x \geq 0) = 0.5$.

To find the probability of a given value of X, it is necessary to compute the corresponding value of x and to refer to the NCEES table.

EXAMPLE 11.4 A continuous random variable X having a normal probability distribution is known to have an arithmetic mean of 14 and a standard deviation of 2.5. What is the probability that, on a given occasion, (*a*) X lies between 14 and 17, (*b*) X lies between 12 and 16.2, and (*c*) X is less than 10?

SOLUTION Refer to Fig. 11.6. This diagram contains two scales: one for X and one for x. The specified range of values is shaded.

Part a. When $X = 14$, $x = 0$. When $X = 17$,

$$x = \frac{17 - 14}{2.5} = 1.20$$

From the NCEES table, $P(x \leq 1.20) = 0.8849$. Therefore,

$$P(0 \leq x \leq 1.20) = 0.8849 - 0.5 = 0.3849$$

(Alternatively, this result can be obtained by taking the value in the last column of the table corresponding to $x = 1.20$ and dividing by 2.) Then

$$P(14 \leq X \leq 17) = 0.3849$$

Part b. Resolve the interval into two parts by cutting it at $X = 14$. For the first part,

$$x = \frac{12 - 14}{2.5} = -0.80$$

From the NCEES table, the area to the left of $x = 0.80$ is 0.7881. Since the area to the left of $x = 0$ is 0.5, it follows that the area between $x = 0$ and $x = -0.80$ is $0.7881 - 0.5 = 0.2881$. (Alternatively, this result can be obtained by taking the value in the last column of the table corresponding to $x = 0.80$ and dividing by 2.) For the second part of the interval,

$$x = \frac{16.2 - 14}{2.5} = 0.88$$

By linear interpolation, the area to the left of $x = 0.88$ is 0.8103. Therefore, the area between $x = 0$ and $x = 0.88$ is $0.8103 - 0.5 = 0.3103$. Then

$$P(-0.80 \leq x \leq 0.88) = 0.2881 + 0.3103 = 0.5984$$

and

$$P(12 \leq X \leq 16.2) = 0.5984$$

Alternatively, the shaded area in Fig. 11.6*b* can be found in this manner: From the table, the area to the *right* of $x = 0.80$ is 0.2119. By symmetry, the area to the *left*

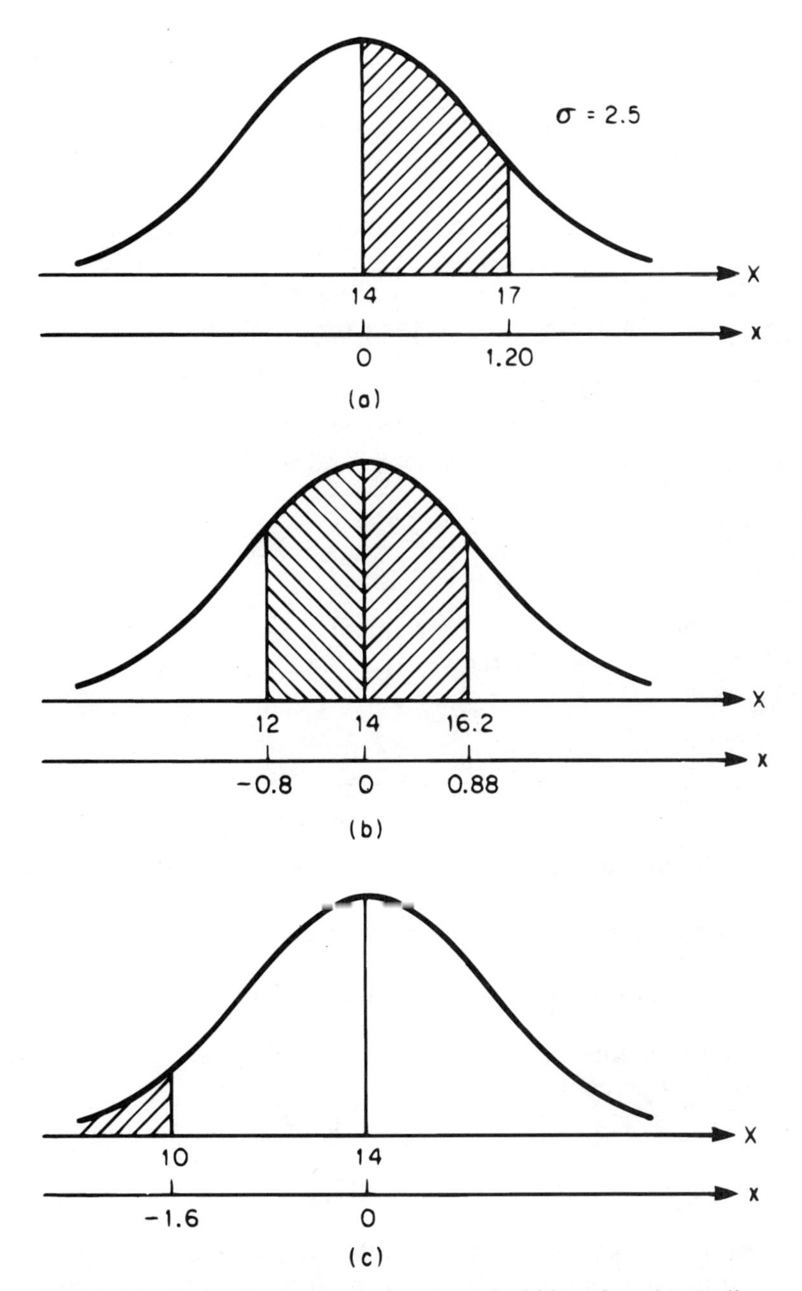

FIGURE 11.6 Determination of probability that (*a*) X lies between 14 and 17, (*b*) X lies between 12 and 16.2, and (*c*) X is less than 10.

of $x = -0.80$ has the same value. As before, the area to the left of $x = 0.88$ is 0.8103. Therefore, the shaded area is $0.8103 - 0.2119 = 0.5984$.

Part c. When $X = 10$,

$$x = \frac{10 - 14}{2.5} = -1.60$$

From the table, the area to the *right* of $x = 1.60$ is 0.0548. By symmetry, the area to the *left* of $x = -1.60$ has the same value, and it follows that

$$P(X < 10) = 0.0548$$

In theory, a random variable can have a normal distribution only if its range of possible values is unrestricted. In practice, however, many variables having a restricted range of possible values are also considered to have a normal distribution if the resulting error is negligible.

EXAMPLE 11.5 A firm manufactures cylindrical machine parts. The diameter of the part is assumed to be normally distributed with a mean of 8.350 cm and a standard deviation of 0.093 cm. A part is considered satisfactory if its diameter lies between 8.205 cm and 8.490 cm. What is the proportion of defective parts?

SOLUTION For illustrative purposes, we have exaggerated both the range of values of the diameter (as manifested in the standard deviation) and the range of acceptable values. Let X denote the diameter, and set $X = 8.205$ cm. Then

$$x = \frac{8.205 - 8.350}{0.093} = -1.559$$

By linear interpolation, the area to the right of $x = 1.559$ is 0.0597, and that is the area to the left of $x = -1.559$. Therefore, $P(X < 8.205) = 0.0597$.

Now set $X = 8.490$. Then

$$x = \frac{8.490 - 8.350}{0.093} = 1.505$$

By linear interpolation, the area to the right of $x = 1.505$ is 0.0662. Therefore, $P(X > 8.490) = 0.0662$.

By summation, the probability that a part selected at random is defective is

$$P(\text{defective}) = 0.0597 + 0.0662 = 0.1259$$

Since probability represents relative frequency in the long run, the proportion of defective parts is 12.59 percent.

11.2.4 Negative-Exponential Distribution

A random variable X is said to have a *negative-exponential* (or simply *exponential*) probability distribution if its probability-density function $f(X)$ is of this form:

$$\begin{aligned} f(X) &= ae^{-aX} && \text{if } X \geq 0 \\ f(X) &= 0 && \text{if } X < 0 \end{aligned} \tag{11.10}$$

where a is a positive constant and e is the quantity defined in Art. 1.20. The probability curve of X appears in Fig. 11.7a. Many mechanical and electronic devices have negative-exponential life spans.

Let K denote a positive constant. By integrating $f(X)\,dX$ between the limits of 0 and K, we obtain

$$P(X \leq K) = 1 - e^{-aK} \tag{11.11a}$$

Then

$$P(X > K) = e^{-aK} \tag{11.11b}$$

Equation (11.11a) expresses *cumulative probability*, and the graph of this equation appears in Fig. 11.7b.

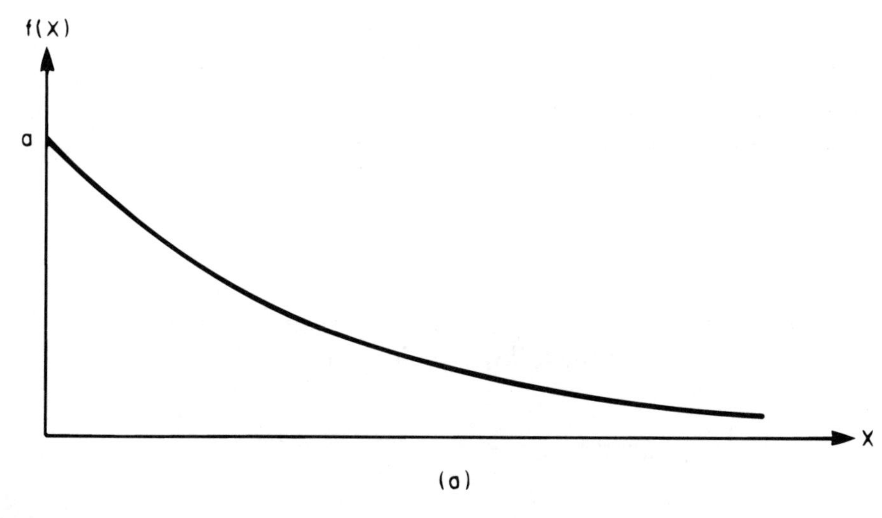

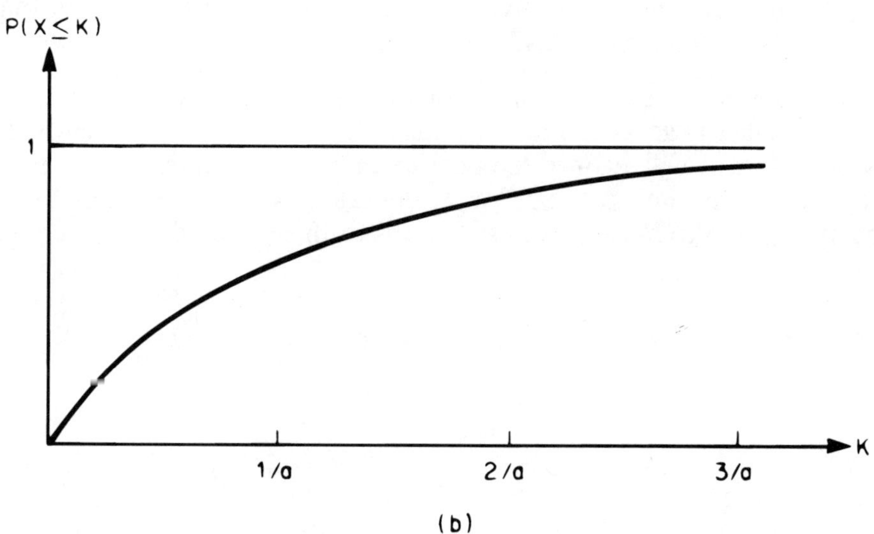

FIGURE 11.7 Negative-exponential probability distribution. (*a*) Probability curve; (*b*) cumulative-probability curve.

The arithmetic mean and standard deviation of a negative-exponential distribution are as follows:

$$\mu = \sigma = \frac{1}{a} \tag{11.12}$$

EXAMPLE 11.6 The life span of a mechanism that operates continuously has a negative-exponential distribution, and the mean life span is 8 days.

a. What is the probability that the life span will exceed 9 days?

b. If the mechanism has been in operation for the past 5 days, what is the probability that it will still be operating 9 days hence?

SOLUTION Let X denote the life span in days. By Eq. (11.12), $a = 1/8 = 0.125$.

Part a. By Eq. (11.11*b*),

$$P(X > 9) = e^{-(0.125)9} = 0.3247$$

Part b. The required probability can be found by applying Bayes' theorem, but we shall follow a different approach by viewing probability as relative frequency in the long run. Assume that a vast number of such mechanisms are activated simultaneously. By Eq. (11.11*b*), the proportion that survive a given interval is as follows:

First 5 days: $e^{-(0.125)5}$

First 14 days: $e^{-(0.125)14}$

Now consider the mechanisms that have survived the first 5 days. The proportion of these that will survive the following 9 days is

$$\frac{e^{-(0.125)14}}{e^{-(0.125)5}} = e^{-(0.125)9} = 0.3247$$

Therefore, if a mechanism has survived the first 5 days, the probability that it will survive the following 9 days is 0.3247.

The values obtained in Parts *a* and *b* of Example 11.6 are equal, and it is apparent that this equality is general. If a mechanism has a negative-exponential life span, the probability that it will survive the next *m* days is independent of its present life. Thus, the mechanism does not "age" as it operates. It ultimately fails because it is continuously exposed to hazards and succumbs to these hazards at some point.

CHAPTER 12
VECTOR ANALYSIS

12.1 DEFINITIONS AND NOTATION

As stated in Art. 9.2.1, a quantity that has magnitude only is termed a *scalar*, and one that has both magnitude and direction is termed a *vector*. Thus, the temperature of a body is a scalar; a force, a velocity, and an acceleration are vectors. The direction of a vector has two characteristics: *inclination* and *sense* (e.g., southwestward). Vectors that are parallel to each other have the same inclination, but their senses may be alike or opposite to each other.

A vector is represented by an arrow. The length of the arrow equals the magnitude of the vector as based on a convenient scale. The arrow is given the same inclination as the vector, and its arrowhead shows the sense of the vector. The term *vector* as used hereafter will refer to the arrow that represents the vector. Thus, we shall speak of the length of a vector rather than its magnitude.

If we are concerned solely with the length and direction of a vector and not its precise location in space, the vector is described as *free*, and we shall deal exclusively with free vectors in this material. It follows that two vectors are equal to each other if their lengths and directions are identical, even though they are at different locations. In our discussion, we shall assume for convenience that the vectors under consideration have a common initial point.

When a letter denotes a vector rather than a scalar, the letter is shown in roman boldface type. Therefore, in the subsequent material, a letter that appears in italics denotes a scalar, and a letter that appears in roman boldface denotes a vector. If A is the initial point and B the terminal point of a vector, the vector is often denoted by $\overrightarrow{AB}$, but we shall use the notation $(\mathbf{AB})$. The length of a vector $\mathbf{v}$ is denoted by $|\mathbf{v}|$ or v; we shall have occasion to use both designations. A vector having zero length is termed a *null* or *zero vector*, and it is denoted by $\mathbf{0}$.

The *negative* of vector $\mathbf{v}$ is a vector that has these characteristics: It is parallel to $\mathbf{v}$; its length in absolute value equals that of $\mathbf{v}$; its sense is opposite to that of $\mathbf{v}$. Thus, if $\mathbf{v}$ is southeastward, $-\mathbf{v}$ is northwestward. It follows that $-(\mathbf{AB}) = (\mathbf{BA})$. Vectors that are perpendicular to one another are called *orthogonal vectors*.

12.2 PRODUCT OF SCALAR AND VECTOR

Consider that a vector $\mathbf{v}$ is multiplied by a scalar s. The product is denoted by $s\mathbf{v}$. Let $\mathbf{w} = s\mathbf{v}$. Vector $\mathbf{w}$ is parallel to $\mathbf{v}$, and its length is s times that of $\mathbf{v}$. The sense of $\mathbf{w}$ is the same as that of $\mathbf{v}$ if s is positive and opposite to that of $\mathbf{v}$ if s is negative.

Thus, the statement $\mathbf{w} = s\mathbf{v}$ expresses both the length and direction of $\mathbf{w}$. It follows that $-7\mathbf{v}$ is a vector that is parallel to $\mathbf{v}$, has a length 7 times that of $\mathbf{v}$ in absolute value, and whose sense is opposite to that of $\mathbf{v}$.

12.3 UNIT VECTORS

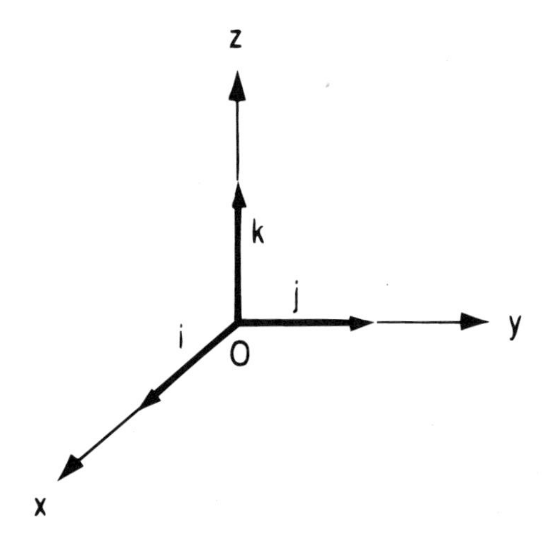

FIGURE 12.1 Basic unit vectors.

A *unit vector* is a vector that has a length of 1 unit. The unit vectors that have the same direction as the x, y, and z axes are denoted by **i**, **j**, and **k**, respectively, and they are referred to as the *basic unit vectors.*

In vector analysis, the coordinate axes are assigned the positions shown in Fig. 12.1. This diagram also displays the basic unit vectors, and these have been placed on their respective coordinate axes for convenience.

12.4 ADDITION AND SUBTRACTION OF VECTORS

Some texts represent the addition and subtraction of vectors by the symbols $\nrightarrow$ and $\rightarrow$, respectively. However, we shall use simply plus and minus signs for this purpose. Let $\mathbf{v}_3 = \mathbf{v}_1 + \mathbf{v}_2$. Vector $\mathbf{v}_3$ is found graphically by the following procedure: In Fig. 12.2*a*, place $\mathbf{v}_1$ and $\mathbf{v}_2$ in the positions shown, with $\mathbf{v}_2$ following $\mathbf{v}_1$. Now draw a vector to close the chain, the initial point of the closing vector being the initial point of $\mathbf{v}_1$ and the terminal point of the closing vector being the terminal point of $\mathbf{v}_2$. The closing vector is $\mathbf{v}_3$.

The foregoing procedure can be extended to the addition of numerous vectors. Thus, let $\mathbf{v}_7 = \mathbf{v}_4 + \mathbf{v}_5 + \mathbf{v}_6$. The addition is performed graphicaly in Fig. 12.2*b*.

Now let $\mathbf{v}_8 = \mathbf{v}_1 - \mathbf{v}_2$, where $\mathbf{v}_1$ and $\mathbf{v}_2$ are the vectors shown in Fig. 12.2*a*. Then $\mathbf{v}_8 = \mathbf{v}_1 + (-\mathbf{v}_2)$, and $\mathbf{v}_8$ is obtained in the manner shown in Fig. 12.2*c*.

The following laws pertaining to vector addition are immediately evident:

$$\mathbf{v}_1 + \mathbf{v}_2 = \mathbf{v}_2 + \mathbf{v}_1$$

$$\mathbf{v}_1 + (\mathbf{v}_2 + \mathbf{v}_3) = (\mathbf{v}_1 + \mathbf{v}_2) + \mathbf{v}_3$$

$$a(\mathbf{v}_1 + \mathbf{v}_2) = a\mathbf{v}_1 + a\mathbf{v}_2$$

$$(a + b)\mathbf{v} = a\mathbf{v} + b\mathbf{v}$$

The first law states that vector addition is *commutative*, the second law states that it is *associative*, and the third and fourth laws state that the multiplication of scalars

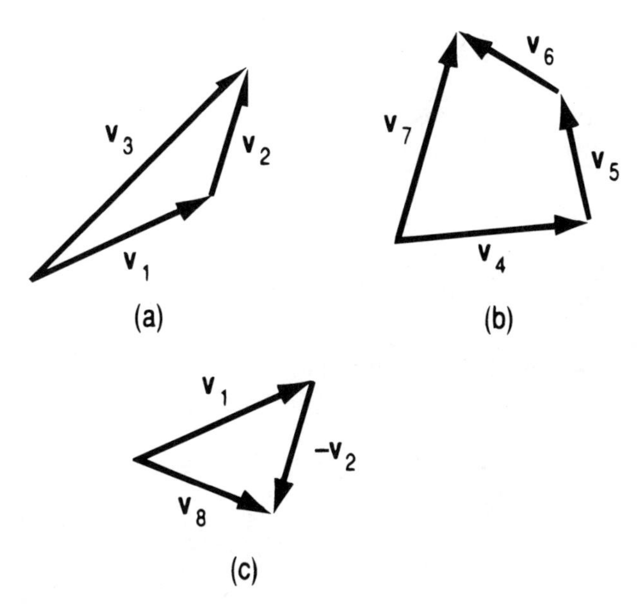

FIGURE 12.2 (*a*) Addition of two vectors; (*b*) addition of three vectors; (*c*) subtraction of vectors.

and vectors is *distributive*. By combining the third and fourth laws, we obtain

$$(a + b)(\mathbf{v}_1 + \mathbf{v}_2) = a\mathbf{v}_1 + a\mathbf{v}_2 + b\mathbf{v}_1 + b\mathbf{v}_2$$

We also have the following: If $\mathbf{v}_1 + \mathbf{v}_2 = \mathbf{v}_3$, then $\mathbf{v}_1 = \mathbf{v}_3 - \mathbf{v}_2$, as can readily be seen in Fig. 12.2*a*. Thus, in a vector equation, quantities can be transposed in the same manner as in an algebraic equation.

With reference to Fig. 12.2*a*, $\mathbf{v}_1$ and $\mathbf{v}_2$ are the *components* of $\mathbf{v}_3$ in their respective directions, and the process of adding vectors is called the *composition* of vectors. Conversely, it is possible to replace a given vector $\mathbf{v}_3$ with two vectors $\mathbf{v}_1$ and $\mathbf{v}_2$ such that $\mathbf{v}_1 + \mathbf{v}_2 = \mathbf{v}_3$ and $\mathbf{v}_1$ and $\mathbf{v}_2$ have specified inclinations. The latter process is called the *resolution* of the given vector.

12.5 BASIC COMPONENTS OF VECTORS

A vector can be described by expressing it as the sum of vectors parallel to the coordinate axes. As an illustration, let $\mathbf{v} = 5\mathbf{i} - 3\mathbf{j} + 9\mathbf{k}$, and let A and B denote the initial and terminal points of $\mathbf{v}$. The vector is constructed in this manner: Starting at A, move 5 units parallel to the x axis and in the positive sense, then move 3 units parallel to the y axis and in the negative sense, and then move 9 units parallel to the z axis and in the positive sense. The point thus reached is B, and $\mathbf{v} = (\mathbf{AB})$.

In general, let

$$\mathbf{v} = v_x\mathbf{i} + v_y\mathbf{j} + v_z\mathbf{k}$$

The scalars v_x, v_y, and v_z are the lengths of $\mathbf{v}$ as projected onto the x, y, and z axes, respectively. Vectors $v_x\mathbf{i}$, $v_y\mathbf{j}$, and $v_z\mathbf{k}$ are the *basic components* of $\mathbf{v}$. The length of $\mathbf{v}$ is

$$v = \sqrt{v_x^2 + v_y^2 + v_z^2} \qquad (12.1)$$

In accordance with the definitions presented in Art. 7.3.3, v_x, v_y, and v_z are *direction components* of **v**, and v_x/v, v_y/v, and v_z/v are the *direction cosines* of **v** with reference to the x, y, and z axes, respectively. Two vectors are equal to each other if and only if the basic components of one are equal to the corresponding components of the other.

EXAMPLE 12.1 Vector $\mathbf{v} = 4\mathbf{i} + 3\mathbf{j} - 12\mathbf{k}$ passes through the origin. Compute the angles that **v** makes with the positive sides of the coordinate axes.

SOLUTION Let θ_x, θ_y, and θ_z denote the angles that **v** makes with the positive sides of the x, y, and z axes, respectively. By Eq. (12.1),

$$v = \sqrt{4^2 + 3^2 + (-12)^2} = 13$$

$$\theta_x = \arccos 4/13 = 72.08^\circ \qquad \theta_y = \arccos 3/13 = 76.66^\circ$$

$$\theta_z = \arccos (-12)/13 = 157.38^\circ$$

EXAMPLE 12.2 Vector **v** starts at $P_1(2,-5,9)$ and terminates at $P_2(13,7,3)$. Express **v** in terms of its basic components.

SOLUTION Let $a_x\mathbf{i}$, $a_y\mathbf{j}$, and $a_z\mathbf{k}$ denote the components.

$$a_x = 13 - 2 = 11 \qquad a_y = 7 - (-5) = 12 \qquad a_z = 3 - 9 = -6$$

$$\mathbf{v} = 11\mathbf{i} + 12\mathbf{j} - 6\mathbf{k}$$

EXAMPLE 12.3 Vector **v** starts at the origin and extends to the midpoint of the line connecting the points $P(15,-7,9)$ and $Q(11,13,5)$. Express **v** in terms of its basic components.

SOLUTION The coordinates of the midpoint are as follows: $x = (15 + 11)/2 = 13$; $y = (-7 + 13)/2 = 3$; $z = (9 + 5)/2 = 7$. Then

$$\mathbf{v} = 13\mathbf{i} + 3\mathbf{j} + 7\mathbf{k}$$

EXAMPLE 12.4 Vector **v** is tangent to the parabola $y = -(3/32)x^2 + 7$ at the point where $x = 4$. If the length of **v** is 20 and its x component is positive, express **v** in terms of its basic components.

SOLUTION The slope of **v** is $dy/dx = -2(3/32)x = -(3/16)4 = -3/4$. The slope of **v** is also the ratio of its y component to its x component; therefore, $a_y = -(3/4)a_x$. We thus have a 3-4-5 right triangle, and $a_x = (4/5)20 = 16$, $a_y = -(3/5)20 = -12$. Then $\mathbf{v} = 16\mathbf{i} - 12\mathbf{j}$.

As Fig. 12.2 discloses, two vectors can be added or subtracted by adding or subtracting, respectively, their corresponding basic components. Thus, let

$$\mathbf{A} = a_x\mathbf{i} + a_y\mathbf{j} + a_z\mathbf{k} \qquad \mathbf{B} = b_x\mathbf{i} + b_y\mathbf{j} + b_z\mathbf{k}$$

Then $$\mathbf{A} + \mathbf{B} = (a_x + b_x)\mathbf{i} + (a_y + b_y)\mathbf{j} + (a_z + b_z)\mathbf{k}$$

and $$\mathbf{A} - \mathbf{B} = (a_x - b_x)\mathbf{i} + (a_y - b_y)\mathbf{j} + (a_z - b_z)\mathbf{k}$$

12.6 SCALAR (DOT) PRODUCT OF VECTORS

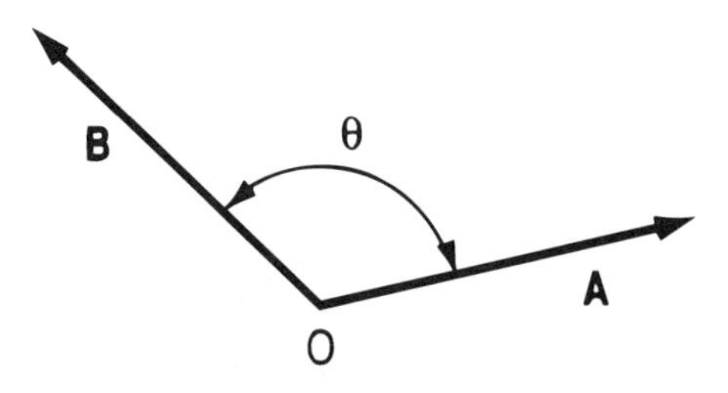

FIGURE 12.3

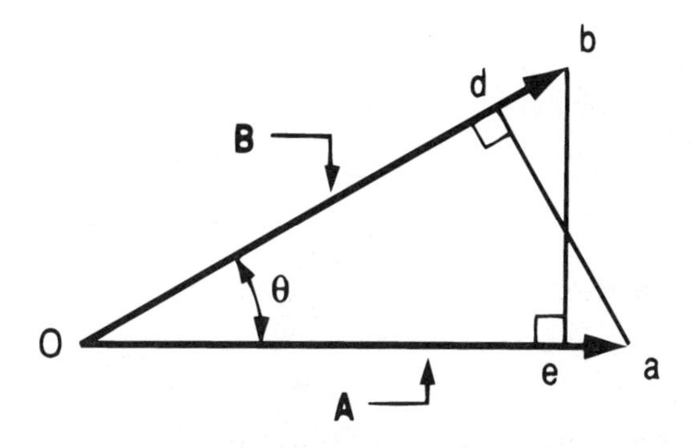

FIGURE 12.4 Interpretation of scalar product of vectors.

Let **A** and **B** denote two vectors, and let θ denote the angle between them, as shown in Fig. 12.3. The *scalar* (or *dot*) *product* of **A** and **B** is denoted by $\mathbf{A} \cdot \mathbf{B}$, and it is defined in this manner:

$$\mathbf{A} \cdot \mathbf{B} = AB \cos \theta \qquad (12.2)$$

As its name indicates, this product is a scalar.

In Fig. 12.4, vectors **A** and **B** have the positions Oa and Ob, respectively. Line ad is perpendicular to Ob, and be is perpendicular to Oa. Then $Od = A \cos \theta$ and $Oe = B \cos \theta$, and it follows that $\mathbf{A} \cdot \mathbf{B} = (Oa)(Oe) = (Ob)(Od)$. Thus, the scalar product can be interpreted as the product of either of the following: the length of **A** and the length of **B** as projected onto **A**; the length of **B** and the length of **A** as projected onto **B**. Manifestly, $\mathbf{A} \cdot \mathbf{B} = \mathbf{B} \cdot \mathbf{A}$, and scalar multiplication is commutative.

EXAMPLE 12.5 Vectors **A** and **B** have a scalar product of -54, $A = 7$, and $B = 12$. What is the angle between the vectors? What is the length of **B** as projected onto **A**? What is the length of **A** as projected onto **B**?

SOLUTION Refer to Fig. 12.5.

$$AB \cos \theta = 7 \times 12 \cos \theta = -54 \qquad \theta = 130.01^\circ$$

$$\text{Length of } \mathbf{B} \text{ as projected onto } \mathbf{A} = \frac{-54}{7} = -7.71 = od$$

$$\text{Length of } \mathbf{A} \text{ as projected onto } \mathbf{B} = \frac{-54}{12} = -4.50 = oc$$

If **A** and **B** are perpendicular to each other, $\cos \theta = 0$, and $\mathbf{A} \cdot \mathbf{B} = 0$. Conversely, if $\mathbf{A} \cdot \mathbf{B} = 0$ and neither vector is a null vector, the vectors are perpendicular to each other. It follows that

$$\mathbf{i} \cdot \mathbf{j} = \mathbf{j} \cdot \mathbf{k} = \mathbf{k} \cdot \mathbf{i} = 0 \qquad (12.3)$$

If **A** and **B** are parallel to each other, $\theta = 0$ and $\cos \theta = 1$. Then $\mathbf{A} \cdot \mathbf{B} = AB$. It follows that

$$\mathbf{i} \cdot \mathbf{i} = \mathbf{j} \cdot \mathbf{j} = \mathbf{k} \cdot \mathbf{k} = 1 \qquad (12.4)$$

It can be demonstrated that

$$\mathbf{A} \cdot (\mathbf{B} + \mathbf{C}) = \mathbf{A} \cdot \mathbf{B} + \mathbf{A} \cdot \mathbf{C} \qquad (12.5)$$

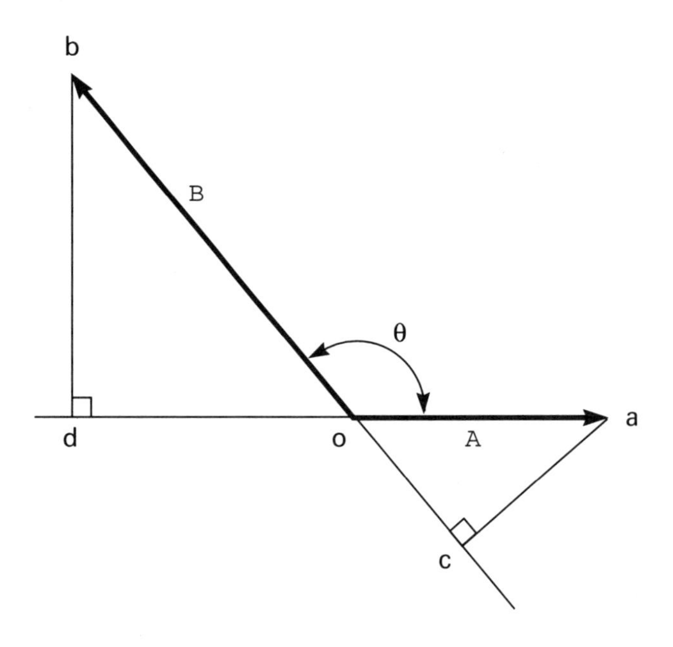

FIGURE 12.5 Vectors and their projections.

Thus, scalar multiplication is distributive. Extending this law, we obtain

$$(\mathbf{A} + \mathbf{B}) \cdot (\mathbf{C} + \mathbf{D}) = \mathbf{A} \cdot \mathbf{C} + \mathbf{A} \cdot \mathbf{D} + \mathbf{B} \cdot \mathbf{C} + \mathbf{B} \cdot \mathbf{D} \tag{12.5a}$$

Now let

$$\mathbf{A} = a_x\mathbf{i} + a_y\mathbf{j} + a_z\mathbf{k} \qquad \mathbf{B} = b_x\mathbf{i} + b_y\mathbf{j} + b_z\mathbf{k}$$

By performing the multiplication and applying the distributive law and Eqs. (12.3) and (12.4), we obtain

$$\mathbf{A} \cdot \mathbf{B} = a_x b_x + a_y b_y + a_z b_z \tag{12.6}$$

This equation enables us to compute the angle between two vectors, as we shall now demonstrate.

EXAMPLE 12.6 Compute the angle between the vectors $\mathbf{A} = -5\mathbf{i} + 8\mathbf{j} - 3\mathbf{k}$ and $\mathbf{B} = 9\mathbf{i} + 6\mathbf{j} - 14\mathbf{k}$.

SOLUTION Equations (12.1) and (12.6) yield the following:

$$A = \sqrt{(-5)^2 + 8^2 + (-3)^2} = 9.8995$$

$$B = \sqrt{9^2 + 6^2 + (-14)^2} = 17.6918$$

$$\mathbf{A} \cdot \mathbf{B} = (-5)9 + 8 \times 6 + (-3)(-14) = 45$$

Since $\mathbf{A} \cdot \mathbf{B} = AB \cos\theta$, we have

$$\cos\theta = \frac{45}{9.8995 \times 17.6918} \qquad \theta = 75.11^\circ$$

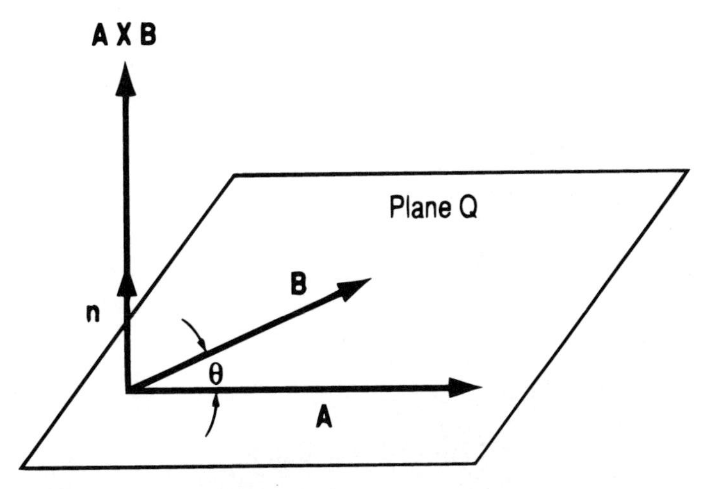

FIGURE 12.6 Definition of vector **n**.

12.7 VECTOR (CROSS) PRODUCT OF VECTORS

In Fig. 12.6, vectors **A** and **B** lie in plane Q, and θ is the angle between them. We impose the restriction $0 < \theta < 180°$. Let **n** denote a unit vector that is normal to Q. The sense of **n** is established in this manner: Consider that **A** is rotated through angle θ to bring it into alignment with **B**. If this rotation is counterclockwise as perceived by an observer, as is true in Fig. 12.6, **n** is directed *toward* the observer; if this rotation is clockwise, **n** is directed *away from* the observer. The sense of **n** can also be established by applying the *right-hand rule*, which states the following: Place the right hand in such position that the fingers point from **A** to **B** in the direction of θ. The sense in which the thumb points is the sense of **n**.

The *vector* (or *cross*) *product* of **A** and **B** is denoted by **A** × **B** and it is defined in this manner:

$$\mathbf{A} \times \mathbf{B} = \mathbf{n}AB \sin\theta \tag{12.7}$$

As its name indicates, this product is a vector, and it is shown in Fig. 12.6. It follows at once that

$$\mathbf{B} \times \mathbf{A} = -\mathbf{n}AB \sin\theta = -(\mathbf{A} \times \mathbf{B})$$

Thus, this form of multiplication is not commutative, and it is imperative that the order of multiplication be clearly specified.

If **A** and **B** are perpendicular to each other, $\sin\theta = 1$, and $\mathbf{A} \times \mathbf{B} = \mathbf{n}AB$. It follows that

$$\mathbf{i} \times \mathbf{j} = \mathbf{k} \qquad \mathbf{j} \times \mathbf{k} = \mathbf{i} \qquad \mathbf{k} \times \mathbf{i} = \mathbf{j} \tag{12.8}$$

If **A** and **B** are parallel to each other, $\sin\theta = 0$, and $\mathbf{A} \times \mathbf{B} = \mathbf{0}$. Conversely, if $\mathbf{A} \times \mathbf{B} = \mathbf{0}$ and neither is a null vector, **A** and **B** are parallel to each other. It follows that

$$\mathbf{i} \times \mathbf{i} = \mathbf{j} \times \mathbf{j} = \mathbf{k} \times \mathbf{k} = \mathbf{0}$$

In Fig. 12.7, we form a parallelogram having **A** and **B** as sides. Let S denote the

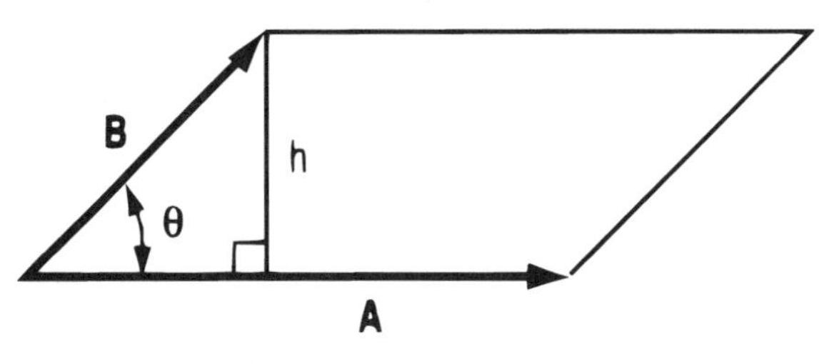

FIGURE 12.7

area of this parallelogram. Then

$$S = hA = (B \sin \theta)A = AB \sin \theta = |\mathbf{A} \times \mathbf{B}|$$

Thus, $|\mathbf{A} \times \mathbf{B}|$ equals the area of the parallelogram having **A** and **B** as sides.

It can be demonstrated that

$$\mathbf{A} \times (\mathbf{B} + \mathbf{C}) = \mathbf{A} \times \mathbf{B} + \mathbf{A} \times \mathbf{C} \tag{12.9a}$$

If we now multiply both sides of this equation by -1, the order of the factors is reversed, and we obtain

$$(\mathbf{B} + \mathbf{C}) \times \mathbf{A} = \mathbf{B} \times \mathbf{A} + \mathbf{C} \times \mathbf{A} \tag{12.9b}$$

An extension of these equations yields

$$(\mathbf{A} + \mathbf{B}) \times (\mathbf{C} + \mathbf{D}) = \mathbf{A} \times \mathbf{C} + \mathbf{A} \times \mathbf{D} + \mathbf{B} \times \mathbf{C} + \mathbf{B} \times \mathbf{D} \tag{12.9c}$$

As before, let

$$\mathbf{A} = a_x\mathbf{i} + a_y\mathbf{j} + a_z\mathbf{k} \qquad \mathbf{B} = b_x\mathbf{i} + b_y\mathbf{j} + b_z\mathbf{k}$$

By performing the multiplication and applying the foregoing relationships, we obtain the following:

$$\mathbf{A} \times \mathbf{B} = (a_y b_z - a_z b_y)\mathbf{i} + (a_z b_x - a_x b_z)\mathbf{j} + (a_x b_y - a_y b_x)\mathbf{k} \tag{12.10}$$

In Art. 3.7, we demonstrated how the determinant of a third-order matrix can be formed. The expression in Eq. (12.10) is recognizable as a determinant, and we have the following:

$$\mathbf{A} \times \mathbf{B} = \begin{vmatrix} \mathbf{i} & \mathbf{j} & \mathbf{k} \\ a_x & a_y & a_z \\ b_x & b_y & b_z \end{vmatrix} \tag{12.10a}$$

EXAMPLE 12.7 If $\mathbf{A} = 3\mathbf{i} - 6\mathbf{j} + 2\mathbf{k}$, $\mathbf{B} = -4\mathbf{i} + 5\mathbf{j} - 9\mathbf{k}$, and $\mathbf{C} = \mathbf{A} \times \mathbf{B}$, identify **C**.

SOLUTION Although the expression for **C** can be obtained by substituting directly in Eq. (12.10), we shall obtain it by applying Eq. (12.10*a*) because this is the one that appears in the NCEES Handbook.

By forming the determinant and then expanding it in the manner prescribed in

Art. 3.7, we obtain the following:

$$\begin{vmatrix} \mathbf{i} & \mathbf{j} & \mathbf{k} & \mathbf{i} & \mathbf{j} \\ 3 & -6 & 2 & 3 & -6 \\ -4 & 5 & -9 & -4 & 5 \end{vmatrix}$$

Forming the products and combining them in the prescribed manner, we now obtain

$$\begin{aligned} \mathbf{C} &= \mathbf{i}[(-6)(-9)] + \mathbf{j}[2(-4)] + \mathbf{k}(3 \times 5) - \mathbf{k}[(-6)(-4)] - \mathbf{i}(2 \times 5) - \mathbf{j}[3(-9)] \\ &= 44\mathbf{i} + 19\mathbf{j} - 9\mathbf{k} \end{aligned}$$

EXAMPLE 12.8 With reference to Example 12.7, vector **D** is perpendicular to both **A** and **B**, and its length is 100. Identify **D**.

SOLUTION Vector **C** in Example 12.7 is perpendicular to both **A** and **B**. By Eq. (12.1),

$$C = \sqrt{44^2 + 19^2 + (-9)^2} = 48.76$$

Then $D/C = 100/48.76 = 2.051$. Since **D** is parallel to **C**, it is simply necessary to multiply the components of **C** by the factor 2.051 to obtain the corresponding components of **D**. However, because the sense of **D** is not specified, two solutions are admissible. They are the following:

$$\mathbf{D} = \pm(90.24\mathbf{i} + 38.97\mathbf{j} - 18.46\mathbf{k})$$

12.8 CALCULUS OF VECTORS

The definition of the derivative of a vector is analogous to that of a scalar. Consider that $\mathbf{v}$ is a function of an independent variable t. Now consider that t increases by an amount Δt, causing $\mathbf{v}$ to increase by an amount $\Delta\mathbf{v}$, which may or may not be collinear with $\mathbf{v}$. The derivative of $\mathbf{v}$ with respect to t is denoted by $d\mathbf{v}/dt$, and its value is

$$\frac{d\mathbf{v}}{dt} = \lim_{\Delta t \to 0} \frac{\Delta\mathbf{v}}{\Delta t} \tag{12.11}$$

For example, if $\mathbf{v}$ represents the velocity of a point, $d\mathbf{v}/dt$ represents its acceleration.

Now let $\mathbf{v} = s\mathbf{u}$, where both s and $\mathbf{u}$ are functions of t. Then

$$\mathbf{v} + \Delta\mathbf{v} = (s + \Delta s)(\mathbf{u} + \Delta\mathbf{u})$$

and it follows that

$$\frac{d}{dt}(s\mathbf{u}) = s\frac{d\mathbf{u}}{dt} + \frac{ds}{dt}\mathbf{u} \tag{12.12}$$

In general, as t increases, both $\mathbf{u}$ and s change. The first term at the right in Eq. (12.12) reveals the effect on $\mathbf{v}$ of the change in $\mathbf{u}$, and the second term reveals the effect on $\mathbf{v}$ of the change in s. The total rate of change of $\mathbf{v}$ is the vector sum of these two vectors.

In the special case where **u** is constant, Eq. (12.12) reduces to the following:

$$\frac{d\mathbf{v}}{dt} = \frac{ds}{dt}\,\mathbf{u} \tag{12.12a}$$

and

$$\frac{d\mathbf{v}}{ds} = \mathbf{u} \tag{12.12b}$$

The derivatives of the scalar product and vector product are as follows:

$$\frac{d}{dt}(\mathbf{v} \cdot \mathbf{w}) = \mathbf{v} \cdot \frac{d\mathbf{w}}{dt} - \frac{d\mathbf{v}}{dt} \cdot \mathbf{w} \tag{12.13}$$

$$\frac{d}{dt}(\mathbf{v} \times \mathbf{w}) = \mathbf{v} \times \frac{d\mathbf{w}}{dt} + \frac{d\mathbf{v}}{dt} \times \mathbf{w} \tag{12.14}$$

In the last equation, it is mandatory that the order of the factors be maintained, since a reversal of the order causes a change in algebraic sign.

Vector integration is simply the inverse of vector differentiation. Thus, if $d\mathbf{w}/dt = \mathbf{v}$, then

$$\int \mathbf{v}\, dt = \mathbf{w} + \mathbf{c}$$

where **c** is an arbitrary constant vector.

Now let $\mathbf{w} = x\mathbf{i} + y\mathbf{j} + z\mathbf{k}$, where the scalars are functions of t. By Eq. (12.12*b*), the partial derivatives of **w** are as follows:

$$\frac{\partial \mathbf{w}}{\partial x} = \mathbf{i} \qquad \frac{\partial \mathbf{w}}{\partial y} = \mathbf{j} \qquad \frac{\partial \mathbf{w}}{\partial z} = \mathbf{k} \tag{12.15}$$

Thus, if y and z remain constant, the increase in w equals the increase in x, and it occurs in the positive x direction.

APPENDIX

DERIVATIVES, INTEGRALS, AND LAPLACE TRANSFORMS

TABLE A.1 DERIVATIVES

Notation: a, c, and n are constants; u, v, and w are functions of x; e is the base of natural logarithms.

Algebraic Functions

$$\frac{dc}{dx} = 0 \tag{1}$$

$$\frac{dx}{dx} = 1 \tag{2}$$

$$\frac{d(cu)}{dx} = c\,\frac{du}{dx} \tag{3}$$

$$\frac{d(u + v + w)}{dx} = \frac{du}{dx} + \frac{dv}{dx} + \frac{dw}{dx} \tag{4a}$$

$$\frac{d(u + v - w)}{dx} = \frac{du}{dx} + \frac{dv}{dx} - \frac{dw}{dx} \tag{4b}$$

$$\frac{d(uv)}{dx} = u\,\frac{dv}{dx} + v\,\frac{du}{dx} \tag{5}$$

$$\frac{d(uvw)}{dx} = uv\,\frac{dw}{dx} + uw\,\frac{dv}{dx} + vw\,\frac{du}{dx} \tag{6}$$

$$\frac{d\left(\dfrac{u}{v}\right)}{dx} = \frac{v\dfrac{du}{dx} - u\dfrac{dv}{dx}}{v^2} \tag{7}$$

$$\frac{d(u^n)}{dx} = nu^{n-1}\frac{du}{dx} \tag{8}$$

$$\frac{d[f(u)]}{dx} = \frac{d[f(u)]}{du}\frac{du}{dx} \tag{9}$$

$$\frac{du}{dx} = \frac{1}{dx/du} \tag{10}$$

Logarithmic and Exponential Functions

$$\frac{d(\log_a u)}{dx} = \frac{1}{u}\log_a e\frac{du}{dx} \tag{11}$$

$$\frac{d(\log_e u)}{dx} = \frac{1}{u}\frac{du}{dx} \tag{12}$$

$$\frac{d(a^u)}{dx} = a^u \log_e a\frac{du}{dx} \tag{13}$$

$$\frac{d(e^u)}{dx} = e^u\frac{du}{dx} \tag{14}$$

$$\frac{d(u^v)}{dx} = vu^{v-1}\frac{du}{dx} + u^v \log_e u\frac{dv}{dx} \tag{15}$$

Trigonometric Functions

$$\frac{d(\sin u)}{dx} = \cos u\frac{du}{dx} \tag{16}$$

$$\frac{d(\cos u)}{dx} = -\sin u\frac{du}{dx} \tag{17}$$

$$\frac{d(\tan u)}{dx} = \sec^2 u\frac{du}{dx} \tag{18}$$

$$\frac{d(\cot u)}{dx} = -\csc^2 u\frac{du}{dx} \tag{19}$$

$$\frac{d(\sec u)}{dx} = \sec u \tan u\frac{du}{dx} \tag{20}$$

$$\frac{d(\csc u)}{dx} = -\csc u \cot u \frac{du}{dx} \tag{21}$$

Inverse Trigonometric Functions

Where a derivative contains a square root, this quantity is to be taken as positive if the indicated angle lies within the following range:

$$-\frac{\pi}{2} \le \sin^{-1} u \le \frac{\pi}{2} \qquad \text{(first and fourth quadrants)}$$

$$0 \le \cos^{-1} u \le \pi \qquad \text{(first and second quadrants)}$$

$$0 \le \sec^{-1} u < \frac{\pi}{2} \qquad \text{(first quadrant)}$$

or

$$-\pi \le \sec^{-1} u < -\frac{\pi}{2} \qquad \text{(third quadrant)}$$

$$0 < \csc^{-1} u \le \frac{\pi}{2} \qquad \text{(first quadrant)}$$

or

$$-\pi < \csc^{-1} u \le -\frac{\pi}{2} \qquad \text{(third quadrant)}$$

$$\frac{d(\sin^{-1} u)}{dx} = \frac{1}{\sqrt{1-u^2}} \frac{du}{dx} \tag{22}$$

$$\frac{d(\cos^{-1} u)}{dx} = -\frac{1}{\sqrt{1-u^2}} \frac{du}{dx} \tag{23}$$

$$\frac{d(\tan^{-1} u)}{dx} = \frac{1}{1+u^2} \frac{du}{dx} \tag{24}$$

$$\frac{d(\cot^{-1} u)}{dx} = -\frac{1}{1+u^2} \frac{du}{dx} \tag{25}$$

$$\frac{d(\sec^{-1} u)}{dx} = \frac{1}{u\sqrt{u^2-1}} \frac{du}{dx} \tag{26}$$

$$\frac{d(\csc^{-1} u)}{dx} = -\frac{1}{u\sqrt{u^2-1}} \frac{du}{dx} \tag{27}$$

TABLE A.2 INTEGRALS

Notation: u and v are functions of x; e is the base of natural logarithms. The natural logarithm is denoted by "ln."

The constant of integration is omitted.

$$\int df(x) = f(x) \tag{1}$$

$$\int dx = x \tag{2}$$

$$\int a\, f(x)\, dx = a \int f(x)\, dx \tag{3}$$

$$\int (u\, dx \pm v\, dx) = \int u\, dx \pm \int v\, dx \tag{4}$$

$$\int x^m\, dx = \frac{x^{m+1}}{m+1} \qquad \text{when } m \neq -1 \tag{5a}$$

$$\int x^{-1}\, dx = \int \frac{dx}{x} = \ln x \tag{5b}$$

$$\int u\, dv = uv - \int v\, du \tag{6}$$

$$\int \frac{dx}{ax+b} = \frac{1}{a} \ln |ax+b| \tag{7}$$

$$\int \frac{dx}{\sqrt{x}} = 2\sqrt{x} \tag{8}$$

$$\int a^x\, dx = \frac{a^x}{\ln a} \tag{9}$$

$$\int \sin x\, dx = -\cos x \tag{10}$$

$$\int \sin^2 x\, dx = \frac{x}{2} - \frac{\sin 2x}{4} \tag{11}$$

$$\int x \sin x\, dx = \sin x - x \cos x \tag{12}$$

$$\int \cos x\, dx = \sin x \tag{13}$$

$$\int \cos^2 x\, dx = \frac{x}{2} + \frac{\sin 2x}{4} \tag{14}$$

$$\int x \cos x\, dx = \cos x + x \sin x \tag{15}$$

$$\int \sin x \cos x \, dx = \frac{\sin^2 x}{2} \quad \text{or} \quad -\frac{\cos^2 x}{2} \tag{16}$$

$$\int \sin ax \cos bx \, dx = -\frac{\cos(a-b)x}{2(a-b)} - \frac{\cos(a+b)x}{2(a+b)} \quad \text{when } a^2 \neq b^2 \tag{17}$$

$$\int \tan x \, dx = -\ln|\cos x| = \ln|\sec x| \tag{18}$$

$$\int \tan^2 x \, dx = \tan x - x \tag{19}$$

$$\int \cot x \, dx = -\ln|\csc x| = \ln|\sin x| \tag{20}$$

$$\int \cot^2 x \, dx = -\cot x - x \tag{21}$$

$$\int e^{ax} \, dx = \frac{e^{ax}}{a} \tag{22}$$

$$\int xe^{ax} \, dx = \frac{e^{ax}}{a^2}(ax-1) \tag{23}$$

$$\int \ln x \, dx = x \ln x - x \tag{24}$$

$$\int \frac{dx}{a^2 + x^2} = \frac{1}{a} \tan^{-1} \frac{x}{a} \tag{25}$$

$$\int \frac{dx}{ax^2 + c} = \frac{1}{\sqrt{ac}} \tan^{-1}\left(x\sqrt{\frac{a}{c}}\right) \quad \text{when } a > 0, c > 0 \tag{26a}$$

$$\int \frac{dx}{ax^2 + c} = \frac{1}{2\sqrt{-ac}} \ln\left|\frac{\sqrt{c} + x\sqrt{-a}}{\sqrt{c} - x\sqrt{-a}}\right| \quad \text{when } a < 0, c > 0 \tag{26b}$$

$$\int \frac{dx}{ax^2 + bx + c} = \frac{2}{\sqrt{4ac - b^2}} \tan^{-1} \frac{2ax + b}{\sqrt{4ac - b^2}} \quad \text{when } 4ac - b^2 > 0 \tag{27a}$$

$$\int \frac{dx}{ax^2 + bx + c} = \frac{1}{\sqrt{b^2 - 4ac}} \ln\left|\frac{2ax + b - \sqrt{b^2 - 4ac}}{2ax + b + \sqrt{b^2 - 4ac}}\right| \quad \text{when } 4ac - b^2 < 0 \tag{27b}$$

$$\int \frac{dx}{ax^2 + bx + c} = -\frac{2}{2ax + b} \quad \text{when } 4ac - b^2 = 0 \tag{27c}$$

$$\int (a + bx)^n \, dx = \frac{(a + bx)^{n+1}}{b(n+1)} \quad \text{when } n \neq -1 \tag{28}$$

TABLE A.3 LAPLACE TRANSFORMS

$f(t)$	$\mathscr{L}[f(t)]$ or $F(s)$
$\delta(0)$ (unit impulse at $t = 0$)	1
$\delta(0)$ (unit impulse at $t = c$)	e^{-cs}
$u(0)$ (unit step at $t = 0$)	$\frac{1}{s}$
$u(c)$ (unit step at $t = c$)	$\frac{e^{-cs}}{s}$
$t[u(0)]$ (unit ramp at $t = 0$)	$\frac{1}{s^2}$
$e^{-\alpha t}$	$\frac{1}{s + \alpha}$
$1 - e^{-\alpha t}$	$\frac{\alpha}{s(s + \alpha)}$
$te^{-\alpha t}$	$\frac{1}{(s + \alpha)^2}$
$\sin \alpha t$	$\frac{\alpha}{s^2 + \alpha^2}$
$1 - \sin \alpha t$	$\frac{s^2 - s\alpha + \alpha^2}{s(s^2 + \alpha^2)}$
$\cos \alpha t$	$\frac{s}{s^2 + \alpha^2}$
$1 - \cos \alpha t$	$\frac{\alpha^2}{s(s^2 + \alpha^2)}$
$t \sin \alpha t$	$\frac{2s\alpha}{(s^2 + \alpha^2)^2}$
$t \cos \alpha t$	$\frac{s^2 - \alpha^2}{(s^2 + \alpha^2)^2}$
$e^{-\alpha t} \sin \beta t$	$\frac{\beta}{(s + \alpha)^2 + \beta^2}$
$e^{-\alpha t} \cos \beta t$	$\frac{s + \alpha}{(s + \alpha)^2 + \beta^2}$

Index

ABOUT THE AUTHOR

Max Kurtz has been teaching courses to prepare engineers for the PE examination continually since 1961, including both Parts A and B of the exam. In addition, he has taught various other engineering courses since 1959. He is the author of numerous books for McGraw-Hill: *Handbook of Applied Mathematics for Engineers and Scientists, Handbook of Engineering Economics, Calculations for Engineering Economic Analysis, Structural Engineering for Professional Engineers' Examinations, Engineering Economics for Professional Engineers' Examinations,* and *Comprehensive Structural Design Guide.*